Flowering Time, Climate and Genotype

The Adaptation of Agricultural Species to Climate through Flowering Responses

Flowering Time, Climate and Genotype

YVONNE AITKEN

MELBOURNE UNIVERSITY PRESS
1974

First published 1974

Printed in Australia by
John Sands Pty Ltd, Halstead Press Division
Artarmon, New South Wales 2064, for
Melbourne University Press, Carlton, Victoria 3053

U.S.A. and Canada: ISBS Inc., Portland, Oregon 97208
Great Britain, Europe, Middle East, Africa and the Caribbean:
International Book Distributors Ltd, 66 Wood Lane End,
Hemel Hempstead, Hertfordshire HP2 4RG, England

National Library of Australia Cataloguing in Publication data
Aitken, Yvonne, 1911–
Flowering time, climate and genotype: the adaptation of agricultural species to climate through flowering responses/ [by] Yvonne Aitken.—Carlton, Vic.: Melbourne University Press, 1974.
Index.
Bibliography.
ISBN 0 522 84071 X.
1. Plants, Flowering of. 2. Crops and climate.
3. Adaptation (Biology). I. Title.
581.54

To L.S.A., W.M.B. and all others who
have helped in the making of this book

Preface

The fact that the many native species in each region of the world flower at the best time for ripening seed in that type of climate is a remarkable and universal phenomenon of plant geography. Yet we know little of the way in which such adaptation has occurred.

The practical aspect raises many questions: What makes a plant flower when it does? Why does one kind of plant flower at a different time from another? Will an early flowering plant make more leaf and seed than a late one of the same species? Why should the one variety of plant set seed more rapidly in one climate than in another?

This type of question is increasingly asked in the present urgency for more flexible and efficient food production. Whether the country is more or less industrialized, the trend is to look critically at the common crops and to try new ones. These may fail at the first attempt because the varieties are unsuited to the local climate. The failure can be caused by maladjusted flowering. This may be either too slow, as when late-flowering European cereals were introduced in the nineteenth century to Australia, or too fast, as with many varieties of root crops introduced more recently into Alaska. In the first case the season ended too soon to allow full seed development—the season was too short for the plant's growing period. In the second case the root crops changed too quickly to the reproductive state and the food storage was reduced—the plant's growing period was too short for high vegetative yield.

These questions can only be answered when a sufficient body of facts has revealed the principles behind the timing of reproductive events in different plant varieties and in various climates. The need for such research spurred the author to conduct the world-wide experiments from 1963 to 1971, which constitute the core of this volume. They are the first in which the patterns of flowering and seed setting of the same well-known varieties of agricultural species have been critically examined in person in a wide range of climatic zones. The comparative analysis of these varieties in relation to the growing season provides not only the desired principles but also 'standard' varieties of known genotype to use in further studies on other species.

The purpose of this book is to analyse the ways in which inherited differences in rate of reproductive development affect the performance, in different climates, of a number of temperate species. The understanding thus gained should permit (a) the more rapid selection of varieties of these species for a given type of climate, (b) the assessment of other species both temperate and tropical, and (c) a more intensive biochemical attack on the nature of development.

Preface

The book is written primarily for the plant scientist, such as the agronomist or plant breeder, who is concerned with the fitting of development to growing season. The reader needs no biochemical knowledge but it is hoped that both biochemist and geneticist will be stimulated to undertake the urgently needed research on plant development while the general reader will gain in deeper insight.

The text is arranged to give a background of history, plant structure, climate and experimental method (chapters 1–4) against which the principles controlling the time of flowering and seeding are described (chapter 5). This is followed by evidence from the climatic plots (chapters 6–9), and the final chapter discusses the application of these principles.

Acknowledgements

I wish to express special thanks to Emeritus Professor G. W. Leeper, Dr B. R. Davidson and Dr W. M. Hiesey for the encouragement without which this book would not have been written, for the reading of the manuscript after the first draft and for their many helpful suggestions.

Many colleagues, both overseas and in Australia, made possible the climatic studies. In 1963 these included Professor E. J. Britten, now at the University of Queensland, Dr R. E. Green and Y. Matsumura of the University of Hawaii, Drs W. M. Hiesey, M. Nobs and J. Clausen and F. Nicolson of the Department of Plant Biology, Carnegie Institution of Washington in Stanford, California, Professors D. D. Chilcote and W. McGuire of Oregon State University in Corvallis, Oregon, Professor G. M. Simpson and W. Diduck of the University of Saskatchewan in Saskatoon, Canada, Dr J. P. Cooper, Pat Mills, Frances M. Jones and R. Thomas of the Welsh Plant Breeding Station in Aberystwyth, Wales, and Flora H. McDonald of the Agriculture School at the University of Melbourne. In 1965 and 1970 others were Drs A. Kallio and D. Dinkel of the University of Alaska, College, Dr L. L. Tieszen of the U.S. Arctic Research Program, Pt Barrow, Alaska, Dr J. Molina, Postgraduate College of Agriculture, Chapingo and Dr R. A. Fischer, CIMMYT, Mexico. Dr J. F. Jenkin of the Botany School, University of Melbourne, and Dr R. I. L. Smith, British Antarctic Survey, University of Birmingham, England, conducted the plots on Macquarie Island (1968) and on South Georgia Island (1970) respectively.

The overseas work was done while on study leave from the University of Melbourne and was assisted by grants from the American Association of University Women, the Wheat Industry Research Council of Australia, the Wheat Industry Research Committee of Victoria and the H. V. McKay Charitable Trust. I also gratefully acknowledge a grant from the University Publication Fund toward the cost of publishing.

I am very appreciative of the people who helped me in various ways locally. Within the University of Melbourne these include W. J. Dahl for the excellent graphs, C. Aeberli and D. Lucas for their devoted care of plants and phytotron respectively, Dr W. J. Collins for statistics, and Sheila E. Hayres.

The historic review became possible with the help of the Baillieu Library staff, especially Dorothy H. Baker and Lois J. Davey, Diana J. Dyason of History and Philosophy of Science, Dr J. Thompson of Middle Eastern Studies, T. H. Chang of East and Southeast Asian Studies, Sue Pringle of Indian Studies, and Ruth Chalmers of Meteorology.

Acknowledgements

The production of the book involved the staff of Educational Technology and Melbourne University Press, especially Jill McDuff with her careful design of the jacket and layout of illustrations, J. Cook with his skilled photography, and the helpful editor Carol Bram.

Other colleagues included Gwen Hotton and A. Clarke of the Victorian Department of Agriculture, W. Adamietz and D. O. Gaffney of the Commonwealth Bureau of Meteorology. I also wish to thank Kath Pullar and Linda Swift for their patient typing and Dr Eva Eden for her support.

Lastly I would like to acknowledge my debt to Emeritus Professor H. C. Forster and to the late Emeritus Professor Sir Samuel Wadham for first arousing my interest in the way climate affects the flowering of plants.

Contents

Abbreviations

FI	flower or inflorescence initiation
LS	leaf stage
MG	maturity genotype
N.S.	new series
pers. comm.	personal communication
R	developmental response
VN	final number vegetative phytomers on shoot

I Historical Background

1 Earliest Records. *Timing of annual plant events. Genetic variation in development rate.*

2 Advances in Knowledge from the Eighteenth Century. *Timing of annual plant events. Genetic variation in flowering time. Organization of shoot apex. Changing concepts of plant development.*

For of course growth [of plants] leads up to reproduction as the completion of the process. THEOPHRASTUS, 4th century B.C.

To every thing there is a season . . . a time to plant, and a time to pluck up that which is planted . . .
ECCLES. 3:2, *Old Testament*, 3rd century B.C.

Types which under certain conditions are characterized by identical vegetative [growing] periods may be of a different physiological and genetic nature . . . The vegetative period, which to a certain degree reflects genotypic characteristics of the particular variety, at the same time depends to an extraordinary degree on external conditions. VAVILOV, 1957

The relationship of the reproductive development of plants to climate is complex. Only isolated facts could be discovered during the many centuries between the first conscious use of plant variation and the present time when technical aids make full analysis possible. The length of the growing period of a particular crop would certainly have been measured in one way or another in relation to the growing season of a particular locality because the yield was vital to survival. Bad management of the crop, a poor season or the wrong variety could mean death by famine when farming was only at the subsistence level. However, the few facts concerning plant development remained purely local until better transport made wider knowledge possible. The facts began to increase and connect up when interest in natural laws became general and such instruments as the thermometer and the microscope were invented.

1 Earliest Records

Early records of plant development are rare, scattered and fragmentary (Table 1.1). They also count from the invention of writing rather than from

Table 1.1
Ancient records of maturity character in plants

Species	Time	Country	Source
Differences in maturity character within a species			
Barley, 'Shesh' grain	3000 B.C.	Sumer (Iraq)	proverbs, myth of Lahar (Kramer 1961)
Millet (*Panicum miliaceum*)	1000 B.C.	China	*Book of Odes* (in Ho 1969, Legge 1871)
Rice	230 B.C.	China	*Book of Rites* (in Ho 1956)
Rice	350 B.C.	India	Panini's Sanskrit grammar (Agrawala 1963)
Fig	600 B.C.	Babylon (Iraq)	King Marduk's garden list (Contenau 1954)
Wheat and barley	350 B.C.	Greece	*Enquiry into Plants* (Theophrastus)
Importance of time of sowing			
Barley	900 B.C.	Palestine	Gezer Calendar (Thomas 1958)
Millet	1000 B.C.	China	*Book of Odes*
Wheat	800 B.C.	Greece	*Works and Days* (Hesiod)
Wheat	30 B.C.	Italy	*Georgics* (Virgil)
Extension of growing period in cooler climates			
One rice variety	A.D. 993	China	Li T'ao (in Ho 1956)
Wheat and barley	350 B.C.	Greece, Egypt	*Enquiry into Plants* (Theophrastus)
Extension of growing period by irrigation			
Barley	1700 B.C.	Sumer	Farmer's Almanac (Kramer 1963)

the actual times when the facts became the common knowledge of each farming village. Nevertheless we can be sure that two of the skills essential for the success of the first agricultural settlements were the abilities to tell the time of the year and to observe and make use of the genetic variation already present in food plants and animals. Both these skills were learnt in the earlier hunting and food-gathering phase and both were needed in the selection of the best type of plant for the local growing season.

TIMING OF ANNUAL PLANT EVENTS

Small communities learnt to estimate time from the regular movements of sun and stars (Pannekoek) and from the seasonal habits of animals such as migratory birds. First by word of mouth and then in writing, calendars set out appropriate action for the management of food production. Month by month each action was related to a particular event such as the rising or setting of a constellation in relation to that of the sun. For instance in Greece the preparation of the land for sowing in the winter-spring growing season could easily be left too late, with consequent failure of the wheat through drought. The farmer-poet Hesiod (800 B.C.) wrote that plowing should begin when the Pleiades set just before sunrise (in early November, late autumn, Dicks 1970).

> At the time when you hear the cry of the crane going over, that annual voice from high in the clouds . . .
>
> But if you have waited for the winter solstice to plow the divine earth, you will have to squat down to reap, gathering in thin handfuls, down in the dust . . .

In Palestine the oldest inscription in Hebrew is the Gezer Calendar (900 B.C.) which is on a small block of limestone found near Jerusalem in 1908 (Thomas 1958). It lists each month's programme, from which one can see that the barley harvest was in the sixth month after sowing. The oldest farmer's almanac (1700 B.C.) is on a clay tablet from the ancient Sumerian site of Nippur near the Euphrates River in Iraq. However, the knowledge must be far older because the Sumerians irrigated crops on a large scale from about 3000 B.C. The tablet gives detailed instructions for the management of the barley crop, even to the value of an extra watering, which will increase the grain yield if the plants are healthy (Kramer 1963).

GENETIC VARIATION IN DEVELOPMENT RATE

Genetic variation in rate of development, as well as in other characters, is likely to have been present within species of potential food value many hundreds of years before the first attempts at their domestication.

The momentous change from hunting to agriculture began in about 7000 B.C. in small Neolithic communities in several regions of the world. These include the hilly flanks of the 'fertile crescent' in south-western Asia (Anati, Braidwood), the south-east part of the loess highlands in China (Ho 1969), and the small river valleys of the plateaux of Central America (MacNeish).

Process of selection

In each case the wild progenitors of the first domesticated plants such as the cereal grasses in south-west Asia and China and the tropical beans and cucurbits in America could be gathered locally. Each population of a potential food species would already have undergone a strong and lengthy natural selection for the maturity character appropriate to the particular locality. The strength of such selection can still be seen in isolated farming areas in the world, as found by Vavilov (1929) in Afghanistan with the mixtures of weed ryecorn and wheat in the fields at different altitudes. Both were of the same late maturity character in the lower, irrigated areas and both were of the early or spring type in the high mountain valleys.

Nevertheless in each of these cases the populations could have included early or later plants because of very localized conditions. Such mixed populations may occur within a small area. This was found with the plants grown from burs of subterranean clover which were collected in Greece and other Mediterranean countries during the expedition in 1954 of the Food and Agriculture Organization of the United Nations.

One may postulate that in a food species the earlier-ripening forms of a local population are likely to have been noticed by the observant gatherer and collected for the first attempts at growing the crop. The same people would have tried to find the occasional plant with seed that was larger, more edible and less prone to rapid and widespread dispersal. Thus such characters as the easily shattering grain of wild oats and the brittle pods of wild peas and beans would tend to become less common in cultivated plants.

The evidence for the selection for better seed retention during the domestication of food plants is much older than that for the recognition of maturity type within the species, because the genetic shift in the first case is obvious in the reproductive structure. For instance, the change from the easily shattering inflorescence of the wild cereal grasses to one that retains the grain much longer can be seen by the way in which the spikelet breaks from the stalk of the ear. Such changes towards the more domesticated crop were achieved as long ago as 5000 B.C. at Jarmo in Iraq in both the emmer and the einkorn wheats and in the two-rowed barley (Helbaek 1969).

Records of maturity differences

References to maturity character within or between species are hard to authenticate without data on times of sowing and harvest. Proverbs and myths from Sumer (3000 B.C.) contain two of the most ancient ones (Table 1.1). In the myth of Lahar the reference to the 'Shesh' grain of forty and sixty days may be better proof of differences in length of development than the reference in the proverbs which mentions 'the early and the late barley' without giving the time of sowing. Two varieties of millet with different growing periods are referred to in the ancient Chinese *Book of Odes*, the *She King* (1000 B.C.). Differences in maturity type are also recorded for other species, including the fig in Babylon in King Marduk's garden list (600 B.C.), rice in India (350 B.C.) and wheat in Greece at about the same time. The Indian writer Panini in his Sanskrit grammar was one of the first to mention the length of time taken by different varieties of a known crop when he compared the 'sixty day' rice, used in the Punjab because it needed little water, with later varieties.

The fact that climate could alter the growing period of a variety became obvious when travelling between countries became easier. When improved farming led to reliable yearly surpluses of food, the isolated villages changed to more complex communities from which empires arose. Officials had to travel widely on government business and the observant ones could note differences in the growing period of the one crop although the variety might not be the same. Thus, in the days of Alexander the Great, barley was seen to take seven months to reach harvest in Greece compared with six in Egypt (Theophrastus). In another case climatic differences are not confused with genetic ones. The success of a military colonization in China in A.D. 993 was based on the high yield of an early rice variety. This was noted to mature a month later than in its original place (Ho 1956).

The more complex societies also made possible the leisure for centres of learning such as the schools of philosophers (600–200 B.C.) in Greece. In the city-state of Athens the philosopher Theophrastus (372–288 B.C.) was intensely interested in plant variation. The phenomenon of development intrigued him, both within and between countries and also in relation to temperature.

He surveyed the wide range of different kinds of plants that flowered successively in the growing season of his locality. Some flowered earlier than others, not only within the group of perennial plants like dandelion and crocus or within the group of annual plants like chick-pea and wheat, but also within the groups of varieties of the same crop such as wheat. He noted that some crops developed fast as shown by the length of time from sowing to ripe seed. Chick-pea was outstandingly fast compared with wheat. Some wheat varieties were of the 'three months' type when grown in summer in favourable situations while others took much longer and grew best when sown in autumn.

He also saw that temperature affected development, although his perception was limited to young plants. '. . . germination begins earlier in sunny places which have an even temperature.' 'In Hellas some are used to sow everything earlier because of the coldness of the soil . . . the object being that the winter may not overtake the crop while it is still tender.' In contrast he noted carefully that the barley and wheat crops in Egypt had shorter growing periods than in Greece but did not implicate the difference in temperature of the growing seasons.

Such was the respect in which Theophrastus was held that his *Enquiry into Plants* remained a text for agricultural writers in Europe for many generations, as can be seen in *The Georgics* by Virgil (30 B.C.) in Italy. However, this same respect may even have helped to stultify further academic interest in plant development until the momentous change from the use of philosophical reasoning alone to the additional use of scientific method in the solving of problems.

Thus the obvious facts of crop development in each countryside continued to be only of local concern until interest in natural phenomena became much more widespread in Europe from about the seventeenth century when various people began inventing better techniques for making instruments that would measure both the environment and the plant (Singer 1959). Fahrenheit in 1715 laid the foundation for accurate temperature measurements by inventing the mercury thermometer. The study of plant development at cellular level only became possible when Galileo in 1609 invented the microscope. However, the publishing of works by great microscopists such as Robert Hooke's *Micrographia* in 1665 was needed to rouse widespread interest. In the late nineteenth century Carl Zeiss and Ernst Abbe helped to refine the instrument. With its aid other workers have greatly extended the knowledge of developmental plant structure.

2 Advances in Knowledge from the Eighteenth Century

Since the eighteenth century knowledge has advanced on four fronts: (a) the timing of plant development in successive years at the same place and in different places, (b) genetic variation in time of flowering, (c) organization of the shoot apex and (d) concepts of plant development.

Flowering Time, Climate and Genotype

TIMING OF ANNUAL PLANT EVENTS

Linnaeus (1751) was the first to propose a systematic recording of seasonal events of plant development (phenology) both within and between countries as a means to show 'how one climate differs from another'. Such a 'regular course of observations, in the manner of astronomers' would be 'of far greater utility to the public'! He also noted that climate affected the growing period of 'the same sort of barley' in Sweden compared with Finland.

Floral calendar

In North America Linnaeus's idea of a floral calendar inspired many. Bigelow (1818) organized observers to note the time of flowering of common fruit trees in the one year in order to measure 'the forwardness of the spring'. Latitude was found to be very important. Peach trees began to blossom earliest in Alabama at 32°N and latest in Montreal, Canada, at 45°N, after another seventy days.

Years later the problem of effective pollination in plum orchards raised interest in the time of flowering of the same varieties of the one species in different parts of the country. Waugh (1898) constructed isophanal charts* for the Burbank and Wildgoose plums which he 'respectfully offered to the plum growing public from the Gulf of Mexico to Hudson's Bay'. Although the time of flowering was later in higher latitude—that is, in cooler spring seasons—he noted the paradox that the Burbank variety blossomed in Vermont at a temperature lower than that of its dormant period in Florida (4·4 compared with 11·7°C).

Temperature and phenology

Reaumur (1735) in France had already used his thermometer to establish the relationship of temperature to the dates of harvest of both wheat and grapes near Paris. The fact that a colder spring and summer than usual had delayed the harvest suggested that a definite sum of daily temperature was needed for the ripening stage of the plant. He wondered whether this would also apply to the ripening of the wheat crop in warmer and colder climates than that of western France.

In the following years, probably spurred by the harsh northern winter, biologists searched vigorously for the thermal laws that would explain differences in the timing of plant and animal behaviour. When the time of flowering of the one group of plants of a perennial species in the one place was recorded from year to year, lower temperature in the preceding months was found to be closely related to later flowering. This type of evidence was gathered by a network of people throughout Europe (Abbe 1905, Schnelle 1955) for a wide range of species and for agricultural routines. The dates included many events: the first cuckoo song, the first bud sprouting on the oak and the first falling leaf, the first flower of spring bulbs such as snowdrop and crocus, the heading of winter wheat, and the timing of farm events such as the first plowing in spring and the cutting of hay. One family in Norfolk,

* Isophane, a line connecting places with the same date for a biological event.

England, recorded the same events for 150 years (Margary 1926) and others in Germany gathered data even through World War II (Schnelle 1955).

The long years of careful work established that a 'temperature constant' (the sum of day-degrees above a minimum) for some species could explain the relationship of temperature to their length of growing period within a small climatic range. Even on the continent of North America, Hopkins (1938) found that trends of temperature with altitude, latitude and longitude were general enough to control such events as time of flowering over large areas through a bioclimatic law.*

However, there were annoying discrepancies to these generalizations in various species and between different places. In Europe, for instance, C. Linsser in 1867 (Abbe 1905) compared the dates of first flowering in the one species in a range of latitude from Belgium to Russia. He showed that a smaller sum of day-degrees was needed for the same phase of development in higher latitudes—that is, where the annual sum of positive temperatures was lower. More recently Nuttonson (1957) found the same trend in the Thatcher variety of wheat when sown in places between 45 and 62°N in North America.

GENETIC VARIATION IN FLOWERING TIME

One cause of the discrepancy between temperature and flowering time was that those searching for thermal laws were ignorant of differences within species in rate of development. Those who were well aware of these differences were usually working on local problems of yield. For instance Marié-Davy in 1882 (Abbe 1905) observed fast- and slower-developing varieties in the cereals, grapes and other crops near Paris and also knew that the maturity character was related to the locality of origin. Varieties from climates to the north of France generally flowered earlier in Paris than those from climates to the south.

However, it was not until the twentieth century that three discoveries fully established the importance of genetic variation. Hence it was realized that the widespread species used for phenological work could have been represented by different maturity characters according to climate.

Firstly Vavilov (1922) found that maturity character varied within each of various cultivated annual species, and formalized this character with many others within his 'law of homologous series in variation'. He also noted that regional groups of varieties within and between species had a similar developmental pattern. At about the same time Turesson (1922) showed in certain cosmopolitan annual and perennial herbs that populations within the same species, when grown together, differed from each other in time of flowering according to habitat of origin.

The classic work of the Carnegie Institution of Washington at the transplant stations in California (Clausen 1940, 1958a) demonstrated the selective power of growing season on maturity character in other perennial species.

* In North America, based from Virginia, an event such as flowering time of a species occurs four days later for every 1° change north in latitude, every 5° change west in longitude and every 100 m increase in elevation.

In a number of native species of different families the alpine varieties, for example, matured earlier than those from lower altitudes. This relationship could be understood when the crossing of climatic races of *Potentilla glandulosa* revealed that complex systems of genes controlled physiological differences such as time of flowering. Thus the several varieties found in the one locality, if crossed, might segregate transgressively and provide fresh variation for local selection.

The other two discoveries concerned the nature of the maturity character. Gassner (1918) found that low temperature hastened the development of 'winter' varieties of cereals and of various biennials. At about the same time Tournois (1914) and Garner (1920, 1923) established the remarkable fact that photoperiod was another climatic factor to affect the time of flowering. As with temperature, plants could react in opposite ways. In some (the short-day group), shortening the photoperiod hastened flowering, as in Japanese hops, hemp, soybean and tobacco. In others (the long-day group), the converse occurred, as in the temperate cereals and in radish. Some other plants were not sensitive. Other workers soon found that temperature could interact with the response to photoperiod (Roberts 1939).

As a consequence, modern studies of the climatic control of plant development include careful selection of variety as well as measurement of the local climatic factors. Each variety of a species must be represented in each type of environment by clones (rooted cuttings of the same plant) or by seeds from the same population (Aitken 1955ab, Blair 1970 pers. comm., Cooper 1952, Evans 1931, Hopp 1972, Schnelle 1964, Smith 1967).

ORGANIZATION OF SHOOT APEX

Ignorance of the morphology of the shoot apex during development has long held back the evaluation of temperature and photoperiod as climatic factors.

The fact that the crop plant must pass through the flowering stage before reaching the ripe seed stage was widely realized from the start of farming. The more astute observer would have postulated an earlier stage as stated by Theophrastus when he wrote that the cereal ear must begin forming 'at the same time that the straw grows', well before its flowering time. However, the influence of climatic factors on the two stages up to first flower remained impossible to assess until flower initiation could be observed with the microscope.

Much has been learnt from studies of the anatomy and the function of the shoot apex and of the process of reproductive development. Investigators from Naudin (1844) to Weatherwax (1930) and Abbe (1941) have described both the external and internal appearance of the vegetative apex and its formation of embryonic leaves in species such as daffodil, cauliflower and maize. The differentiation of the reproductive apex has been recorded for various grasses, from Trécul (1880) to Bonnett (1935), Evans (1940) and Sharman (1945). The fact that young flower buds could be found on shoots long before their flowering, was noted by Gray (1881) for some forest species, by Goff (1899) for orchard trees in North America, and also by workers on arctic vegetation, such as Sørensen (1941) in Greenland. Goff observed that 'the expansion of the flowers in spring is little more than the distention of the cells formed in the preceding year'—that is, about ten months after flower initiation! In wheat the problem of yield led to a comparison of the timing of the growth of the young ear within the plant with external signs of development, by Carruthers (1892) and Hudson (1934) in England, and Kiesselbach (1926) in Nebraska in the United States.

In the last fifty years interest in the nature of the maturity character within a species has concentrated more attention on the change to the reproductive state at the shoot apex. Purvis (1934) was one of the first to show the key importance of flower initiation to earliness of flowering in rye. She also found that early and late varieties differed in sensitivity to temperature and photoperiod within the first stage of development. Cooper (1956a) established the fact of genetic variation in both the timing of flower initiation and in the rate of development between flower initiation and flowering in a number of grass species. This fact provides the basis for his technique of developmental analysis of varieties of different habitats in order to understand their responses to climatic factors.

Purvis also showed that the timing of flower initiation provided another index of maturity, the final number of vegetative nodes. Because the rate of leaf appearance is alike for varieties of the same species and only reproductive nodes are formed at the apex after the change to the reproductive state, early flower initiation results in a low number of vegetative nodes. Some years earlier Vavilov (1923) in his exhaustive study of the characters of the bread wheat had noted the relationship between time of flowering and vegetative nodes on the shoot: 'This character is in connection with early maturity; the more there are nodes, the more the race is late-ripening.' A colleague of his, Kuleshov (1931), recorded the same association in maize. Other workers have discovered the relationship themselves, such as Cooper (1956a) in both cereals and herbage grasses, and Aitken (1941) in subterranean clover.

The change to the reproductive state is now also understood to some degree in terms of hormonal balances under genetic control and is capable of being affected by photoperiod and temperature in several ways (Evans 1969, Leopold 1964).

One aspect of the change to the reproductive state is the location of the plant responses to the climatic factors, within certain tissues of the plant. Since responses to photoperiod occur in the leaf (Knott 1934), the 'flowering stimulus' must be made and transported to the apex in sufficient quantity for the change. In spite of favourable photoperiod, drastic leaf removal in the early vegetative stage can retard development to flower initiation in subterranean clover and this may occur when the species is grown in pastures under high stocking rates (Collins 1970). Likewise low nutrient status of the leaf may reduce the supply of the stimulus to the apex, as suggested by the retarding effect on flower initiation of the removal of cotyledons in seedling peas (Collins 1972) and of low nitrogen in cocksfoot (Calder 1961).

With regard to temperature, both positive and negative responses must occur in the same tissue and therefore, in some varieties and situations, at the same time. The positive response to temperature is in the shoot apex and in the adjoining tissues. The response that prevents flowering at a high temperature is localized at the shoot tip, as shown by the fact that low temperature in this region permitted flower initiation in celery and beet varieties, which otherwise remained vegetative in high temperature (Curtis 1930, Thompson 1939).

CHANGING CONCEPTS OF PLANT DEVELOPMENT

The observant individuals within every migrating group of people must have noted that the same variety of plant might develop faster or slower in the new climate than in that of the native land. However, they seldom wrote about these plant events, and the theories of plant development have increased in complexity as facts have accumulated over the last two centuries. The three main concepts that have emerged in turn from the mass of data are as follows:

'Heat unit' theory

The first theory, that a fixed quantity of heat is needed for a plant to reach a particular stage, was proposed by Reaumur in France as early as 1735. This idea stimulated the interest of botanists in temperature efficiencies for plant growth and development (Abbe 1905, Price 1910, Lebenbauer 1916). From the end of the nineteenth century the 'heat unit' theory was applied to the climatic zoning of agricultural crops on a continental scale in North America (Merriam 1894, Livingston 1913, Kincer 1922, Hopkins 1938).

Nowadays the theory has been proved to be only partially useful for prediction of time of flowering when varieties sensitive to photoperiod are grown in a range of latitude. One variety of Timothy grass (Evans 1939) should have flowered 80 days later at 65°N than at 32°N, according to Hopkins's law, but only took 40 days, probably because of its response to photoperiod.

However, the theory is of real value in the vegetable processing industry (Bomalaski 1948, Holmes 1959) with crops in which varieties of simple maturity genotype are grown in an area that differs in temperature through a fairly long growing season. Suitable varieties of crops such as pea, bean (*Phaseolus vulgaris* L.) and maize can be sown at a series of times designed for progressive harvesting on predicted dates of maturity.

Concept of phasic development

The concept assumes that each stage of development in a plant has definite requirements for temperature and/or photoperiod and that these requirements may vary for the different stages. Progress from one stage to the next may be prevented by an unfavourable supply of the necessary factors.

This theory takes a substantial step forward and is largely adequate for some species, including subterranean (Aitken 1955ab) and white clovers (Britten 1961) and annual and perennial ryegrasses (Cooper 1957). These have late-flowering varieties with strong temperature-negative responses in their early development. When sown in spring in the temperate zone, they remain vegetative over summer and until the winter has provided the 'cold requirement' for flower initiation. Similarly one can explain the fact that the late-flowering red clover variety, when sown in late summer in Melbourne, stays vegetative until the following spring, by its requirement for a long photoperiod in the first stage.

The situation is more complex than this for the species for which the theory was proposed. The fact that the late-flowering or winter cereals (Gassner 1918, McKinney 1933, Purvis 1934), when sown in spring, did not flower until the following spring was explained by the absence of sufficiently low temperatures over the summer to complete their vegetative stage. But evidence from these late-flowering varieties shows that flower initiation is only delayed, not prevented, in high temperature and that development fails temporarily early in the reproductive state because of maladjusted responses to the unfavourable combination of temperature and photoperiod over summer and autumn.

In general the evidence supporting the theory of phasic development suggests that maturity character in a species is based on increasing grades of sensitivity to temperature and/or photoperiod, particularly for the change to the reproductive state.

Concept of development analysis

Development analysis as proposed in the present work concentrates on the range of maturity character both within and between species, with the aim of understanding the way each maturity genotype produces a growing period of definite length in a given environment. The maturity genotype is made up of a group of units which include the tendency to flower and a number of responses two of which are universal ones. These units interact with the temperature and photoperiod of the growing seasons by varying the components of development (vegetative nodes, rate of leaf appearance and rate of seed maturation).

One of the first studies with this approach was that of Forster (1935) who examined in the field, with supplementary light, the response of English and Australian wheats to temperature and photoperiod throughout their growing period. Although without full analysis of the components of development, the evidence showed that the two early varieties and the one late variety differed in response to the two climatic factors through the first two stages up to heading. Both the emergence of the seedling and the ripening of heads were controlled by temperature only. Fresh examination of the data shows that the varieties also differed in their tendency to flower. This was best indicated when they grew at low temperature and long photoperiod—that is, when the late variety was not retarded by its temperature-negative response and the three varieties were alike in their responses to photoperiod.

As the author's climatic studies provide many other examples of the relevance of development analysis, further discussion is best postponed to chapter 10, when the reader will be more familiar with the evidence.

II Plant Development and Climate

A comparison of different climates shows how closely the time of flowering of a local flora is adjusted to the most favourable period of the year for seed production. This is seen most dramatically in the species native to arctic and alpine regions. Although growth is possible only within two or three cool summer months, a surprising number of species manage to flower then and even to set seed (Sørensen 1941, Clausen 1940). In the middle latitudes, places with favourable temperature and rainfall confined to the warmer part of the year have most plants flowering in spring and setting seed over summer, a fact widely celebrated by European poets. In contrast drier parts of the countries bordering the Mediterranean Sea or of California and southern Australia have a mild winter but a long, dry summer. Seed setting there occurs before the drought because of flowering in the winter and early spring. In Western Australia, for instance, the rich native flora of the south-western part scatters vivid patches of colour through the bushland at this time. In the warm, moist tropics, conditions are favourable all the time and flowering may occur throughout the year depending on the species.

The key to the close adaptation of flowering plants to climate lies in the fact that inherited differences in the time of flowering and of seed setting (the maturity character) occur within and between species. Thus the problem is to understand the way in which climatic factors affect the timing of these reproductive events—that is, the rate of *reproductive* development. In the context of climate in this book the word 'development' is used in the special sense of progress towards sexual reproduction as distinct from growth (see chapter 3).

1 Genetic Diversity in the Maturity Character

Differences in time of flowering within a species have been established in both wild and cultivated plants. Evidence of natural selection for the appropriate form of this character comes from the transplant studies of ecological races of various native species in California (Clausen 1958ab). The history of agricultural crops provides much more evidence, probably because the watchful farmer under the threat of famine actively selected the most useful maturity character of his valuable crops for seed. For instance, Vavilov (1929) found in Afghanistan that late-flowering wheats and flax were used in the lower country with the longer growing season and early-flowering ones in the high mountain valleys. Similarly, late-flowering varieties of barley were used in the plains of Hungary where the long growing season began in autumn while very early-flowering varieties were typical of those from the short growing season in Egypt (Orlov 1931).

Only within the last hundred years has the genetic diversity in the maturity character been realized, mainly through the efforts of Vavilov (1922), whose law of homologous series emphasized the similarity of genetic variation within each of a number of cultivated plants.

However, even before Vavilov, there was intense interest in the 'newer' countries with their short growing seasons in the possibilities of crop improvement. In Australia cereal growing was improved by using as parents early-flowering varieties from India, South Africa and California (Farrer 1898) while in Alaska the first promising early varieties came from Scandinavia and Siberia (Georgeson 1899).

In the last few years advances in the sciences of genetics, plant physiology and agronomy and also in the technology of seed production and preservation at last make possible a world view of plant resources. The international organization necessary for the supply of genetic stocks for agronomic needs in any country has also been started. The outstanding success of the International Programmes for food crops in raising the yield of wheat and maize in Mexico and rice in the Philippines (Stakman 1967) shows that such a plan is feasible. Nevertheless the plant breeding programme for even these species is still being slowed by ignorance of the maturity character and its relationship to yield in a given climate.

2 Interaction of Maturity Genotype and Climate

Little is known about the control of rate of development by environmental factors, particularly climatic ones, yet a delay of only a few days in the timing of a stage of development can mean complete loss of a crop, as when the seed in some varieties of grain crops in Alaska matures too slowly.

In order to select varieties for a limited growing season, we need a thorough understanding of the genetic factors controlling the growing period (days from sowing to ripe seed) and of the environmental factors that react with them. In particular an early-flowering variety can cope better with a short growing season than a late-flowering variety (Aitken 1955abc, 1966a).

In the past, much of the work on climate and development has been too scattered and without sufficient detail, in spite of the devotion of many people over the last century. More recently several advances have been made. The first was that the importance of observing the one genotype in climatic studies was realized. Cooper (1952) used clones of S24 perennial ryegrass to measure the effect of spring temperature on time of flowering in different localities in Great Britain. Another advance was the increased interest in the critical but obscure change from the vegetative to the early reproductive stage at flower initiation. Cooper (1956a, 1959a) established that the timing of the change in perennial ryegrass was affected by differences between varieties in sensitivity to both temperature and photoperiod and that this was the main reason for the much more obvious differences in time of flowering.

Nowadays as evidence accumulates on the nature of the genetic control of flowering (Evans 1969, Leopold 1964) we also look for individual processes or chemical substances that promote or hinder the rate of development towards flowering. Thus earliness within a species may be divided into two components. The first determines that only a small number of vegetative nodes may be required before flowering begins (Purvis 1934, Gott 1955, Aitken 1955abc, 1961a). The second is that a given experience of temperature or photoperiod may have a greater effect on the change to the reproductive state in one variety than in another. Evans (1959) proposed three processes which affect flower initiation in subterranean clover, two of which were promoted by high temperature and one inhibited. Not only may different responses operate at the same time in the one variety, but the variety may respond differently to the same value of a climatic factor before and after flower initiation (Aitken 1955a).

The stage has now been reached for the analytical approach to the problem of maturity type and climate on a broader front and in greater depth. It is thus essential to collect two kinds of experimental information. The first involves comparing a number of genotypes growing in the same climate over the full span of development; the other, observing the same genotype in a wide range of climate.

The aim of this book is to present the results of this two-fold approach from experiments on some well-known species grown in different parts of the world. The method is outlined below.

The picture is extremely complex and a simplifying system has been used for presentation, namely that the several stages in the plant's development are determined by one or more of a number of individual reactions. The strength of any one is genetically determined and some may act in opposition to others. This type of developmental analysis, as distinct from growth analysis, allows the maturity genotype of a variety to be assessed. Its expression in a given type of climate may then be better understood.

This book is not directly concerned with the chemical nature of the substances affecting the rate of development but provides a great deal of fresh material for biochemist and geneticist to use for this purpose. It is also hoped that the present survey of nine agricultural species will serve as the nucleus for a much more widespread attack on the interaction of maturity genotype (MG) and climate in both temperate and tropical plants.

3 Experimental Method Used in Author's Genotype–Climate Studies 1963–71

In order to study the effect of climate on the development of a plant, it is essential to select plant material that represents a range of maturity genotype

Table 2.1
Species, maturity group and variety used in the study of genotype in relation to growing season in diverse climates

Species and agricultural use	Early maturity	Later maturity
Grain legume		
Pisum sativum L.	Alaska	Dun
pea	Collegian	Partridge
	Canner's Perfection	Mackay
	Mansholts	Austrian Winter
	MU38	
	MU722	
	Frosty	
Cereals		
Avena sativa L., *byzantina* C. Koch*	Early Wonder	Algerian
oat—common, red	Alpha	Golden Rain
	Rapida	Mugga
	Nips	Peniarth
Hordeum vulgare L.	Finlay	Fimbul
barley	Bankuti	Bore
	Noyep	
	Research	
	Cape	
Secale cereale L.	South Australian	Varnerag
rye		Rheidol
Triticum aestivum L.	Sunset	Kharkov
wheat—common or bread	Thatcher	Winglen
	Gabo	Diana
	Spica	Capelle Desprez
	Insignia	Holdfast
	Kogat	Sammet
Pasture legumes and grasses		
Trifolium pratense L.	Cowgrass	Montgomery
red clover		
Trifolium subterraneum L.	Dwalganup	Mt Barker
subterranean (sub) clover		Tallarook
Lolium rigidum Gaud.	Merredin	Wimmera
annual ryegrass		
Lolium perenne L.	Kangaroo Valley	Victorian
perennial ryegrass		

* Stanton 1955

controlling the rate at which the plant reaches the flowering and ripe seed stages. It is also essential to standardize both the measures of development and the culture of the varieties so that no factors other than the climatic ones of temperature and photoperiod can affect rate of development.

VARIETIES

Early and late varieties were chosen from nine species in which the flowering character had been studied (Aitken 1945, Aitken 1955ab, 1961a, 1964, 1966ab). Most of the varieties are used in Australia but some that are not are included, especially within the late-flowering group. The species all have a low threshold of temperature for active growth and are therefore widely used in the temperate parts of the world. They are confined to the herbaceous group and include a grain legume, cereals, and some pasture species

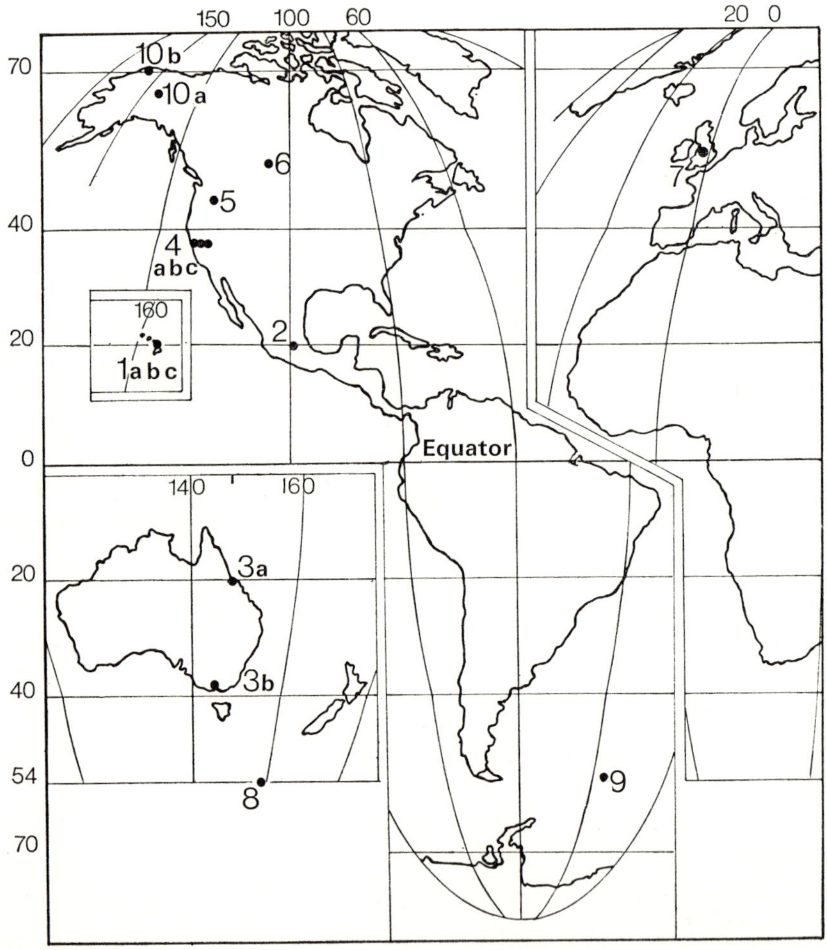

Fig. 2.1 The experimental sites in relation to country and latitude.

(Table 2.1). All of the species are annual except perennial ryegrass and red clover. In four species the varieties represent several grades of earliness and lateness.

METHOD

Three climatic factors—temperature, photoperiod and rainfall—are the main ones to affect plant development but only the first two will be considered here. Various combinations of temperature and photoperiod were obtained from sowing the above varieties at different times in a range of latitudes and altitudes (Fig. 2.1, Table 2.2, Appendix 2).

Table 2.2
Index of experimental sites in relation to country, latitude and altitude

Site index	Symbol used in figures	Location	Latitude degrees	Altitude metres
1a	●	Hawaiian Is—Waimanalo, Oahu	20 N	18
1b	●	Kula, Maui	20 N	1067
1c	●	National Park, Maui	20 N	2042
2	■	Mexico—Chapingo	20 N	2200
3a	◆	Australia—Ayr	20 S	10
3b	◆	Melbourne	38 S	40
4a	▼	California—Stanford	38 N	30
4b	▼	Mather	38 N	1400
4c	▼	Timberline	38 N	3050
5	▲	Oregon—Corvallis	45 N	85
6	◐	Canada—Saskatoon	52 N	515
7	○	Wales—Aberystwyth	52 N	30
8	□	Macquarie I.	54 S	45
9	◇	South Georgia I.	54 S	3
10a	▽	Alaska—College	65 N	153
10b	△	Pt Barrow	71 N	10

Plant culture

In each environment the soil water and fertility were kept favourable for growth and the number of plants of each variety was sufficient for adequate sampling of development. Each variety was sown to give 30–50 plants in an area of 40 by 30 cm. The seeds were sown in four adjoining rows, each 13 cm apart.

Plant selection

In each variety the plants observed for dates of flowering and ripe seed were confined to the two middle rows in order to avoid the possible complications of wider spacing on these measures. Those needed for dissection of the shoot apex and for special marking were selected from any row. For the latter purpose, ten healthy seedlings were marked on the third leaf on the main shoot (see Fig. 3.8) and on each successive third leaf until flowering.

These plants were used to calculate the mean leaf stage of the variety at a given date and the vegetative nodes or phytomers at flowering when difficult to count otherwise.

The observations were made on the main shoot of each plant except in the case of lateral dominance in which the most vigorous lateral was also included. In the grasses, date of heading was taken rather than that of first flower, because recording was easier and more reliable.

4 Types and Definitions of Development Measures

TYPES OF MEASURES

Development was measured in two ways. The first set of measures was related to time as follows:

1. Date of first flower on the main shoot of annual plants such as pea and subterranean clover or of first reproductive shoot in perennial species such as red clover; date of heading in grasses.
2. Number of days from sowing to first flower (pea, subterranean clover, red clover) or to heading (the four cereals and the two ryegrasses).
3. Number of days from sowing to ripe seed.
4. Number of days for each of the three developmental stages: (a) sowing to flower initiation (FI), (b) FI to first flower (pea and clovers) or to heading (cereals and grasses), (c) first flower or heading to ripe seed.

These measures are valuable when the growing period of a variety is considered in relation to the available growing season (the number of days with favourable conditions for growth) of the one environment. However, they may be useless for predicting the development of the variety in a different environment.

The structure of the shoot can provide another set of measures which are complementary to the first set and include:

1. Leaf stage at FI.
2. Total phytomers, units of shoot growth, both mature and immature, at FI.
3. Number of vegetative phytomers (VN) at first flower or heading.

DEFINITIONS OF MEASURES

Growing period: number of days from sowing to ripe seed.

Development stage 1 (vegetative): number of days from sowing to FI. This was calculated by dissecting the shoot tip to note the first appearance of axillary reproductive primordia or 'double ridges'. The date was taken when half the samples had reached the stage.

Development stage 2 (early reproductive): number of days from FI to first flower in legumes or to heading in grasses. The date of this event was taken when half the marked plants had reached the stage of first open flower on the first inflorescence in pea and the clovers or of heads peeping in the grasses—that is, the first appearance of the topmost spikelet above the ligule of the top leaf.

Development stage 3 (late reproductive): the number of days from first flower or heading to ripe seed. The date of this event was taken when half the marked plants had most seeds dry on the main shoot.

Development rate: the inverse value of the number of days between sowing and ripe seed or between the beginning and the ending of a development stage.

Leaf stage (LS): number of mature phytomers on a marked shoot, counted from the first phytomer above the cotyledons or coleoptile as one on the main shoot or as one from the first node at the base of a vigorous lateral. A new leaf stage is attained in legumes when the youngest mature phytomer first opens its leaf or in grasses when the ligule is first seen at the base of the youngest leaf blade.

Leaf stage at flower initiation (FI): number of mature phytomers on the main shoot at FI as shown by the number of fully open leaves.

Plastochron: number of days between the formation of one new leaf primordium (the youngest immature phytomer) and the next, as found by dissection and by counting the number of immature phytomers between the top open leaf and the shoot apex.

Rate of leaf appearance: the inverse value of the *phyllochron,* the number of days between the opening of two successive leaves (see also *Plastochron*).

Vegetative phytomers (VN) at first flower or heading: this measure is equivalent to 'internode number' (McKinney 1933), 'final leaf number' (Purvis 1934) and 'final vegetative nodes'. It is one less than 'node of first flower' in species with axillary inflorescences (Aitken 1941). The number of vegetative phytomers is the total of mature and immature ones below the first reproductive phytomer at FI. The number can be counted at FI but is more easily counted from the number of vegetative nodes at first flower.

III Plant Development and Shoot Structure

The process of development in the green plant is described in this chapter in order to appreciate the concept of maturity genotype (chapter 5) and its expression in different climates (chapters 6–9). Development will be considered in three sections, the first on the definition of the process, the second on the shoot structure and the third on measurements.

1 Process of Development

DEFINITIONS OF GROWTH AND DEVELOPMENT

Although full development cannot occur without some growth, the two processes must be distinguished first in general and then in relation to the shoot structure. Growth and development are concurrent aspects of the plant from its beginning as a single cell. The daily increase in cell number and differentiation increases the dry weight (*growth*), and changes in the hormonal balance direct the progress of the shoot apices from the vegetative to the reproductive stages (*development*).

The plant grows first on maternal food reserves in its progress from zygote to embryo and to seedling. From this stage onwards it depends on the activity of its photosynthetic tissues. Growth is measured as a daily rate (relative growth rate, g/g/dy; absolute growth rate, g/plant/dy) and may be partitioned into that of the shoot as distinct from the root. The dry weight of the plant at a given time is its 'biological' yield (Donald 1962) and is closely related to the amount and activity of photosynthetic tissue.

The development towards sexual reproduction takes place in the shoot, passing through three main stages (Fig. 3.1). The first or *vegetative* stage ends

a Stages of Development

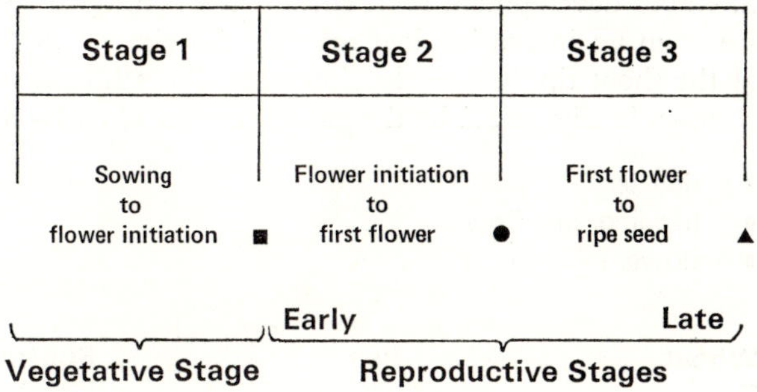

b Components of Development

1 Rate of leaf appearance
2 Number of vegetative nodes } These affect stages 1 and 2
3 Rate of seed development } This affects stage 3

Fig 3.1 (a) The growing period of the shoot of an annual plant is made up of three stages of development. The end of each stage is marked in turn by the developmental event of flower initiation, first flower and ripe seed as shown by the symbols. (b) The length of each stage is affected by one or more components of development.

with flower initiation, the second or *early reproductive* stage ends with the onset of flowering and zygote formation, and the third or *late reproductive* stage ends with the ripening of the seed. The first stage is counted from the date of sowing, but it should not be forgotten that some advance towards the second stage may have already taken place in the embryo, depending on the type of plant and environment.

Development is measured by the length of each stage and of the total growing period. The daily developmental rate is inversely related to the number of days taken to complete a stage. When compared in the one environment, varieties of a species may vary widely in maturity character. Those that develop rapidly in all three stages have a short growing period and are 'early-maturing' while those that develop more slowly are 'later-maturing'. Rapid development of a variety through the first stage is usually closely associated with both early flowering and ripening of seed. However, some forms of stress retard later development if they kill flower buds or developing seed, or prevent seed setting. Thus the variety may be mistakenly assigned to a later-maturing group in an unfavourable growing season although it correctly belongs to an early-maturing group.

RELATIONSHIP OF GROWTH AND DEVELOPMENT

Although growth and development usually occur at the same time in the plant, they differ in two ways. From one aspect, development competes with growth. Rapid development soon curtails the continued formation of photosynthate from an increasing leaf area by the change to reproductive structures at the shoot tip. Slower development allows further vegetative growth. This is shown in Figure 3.2 by the growth curves of early- and later-maturing

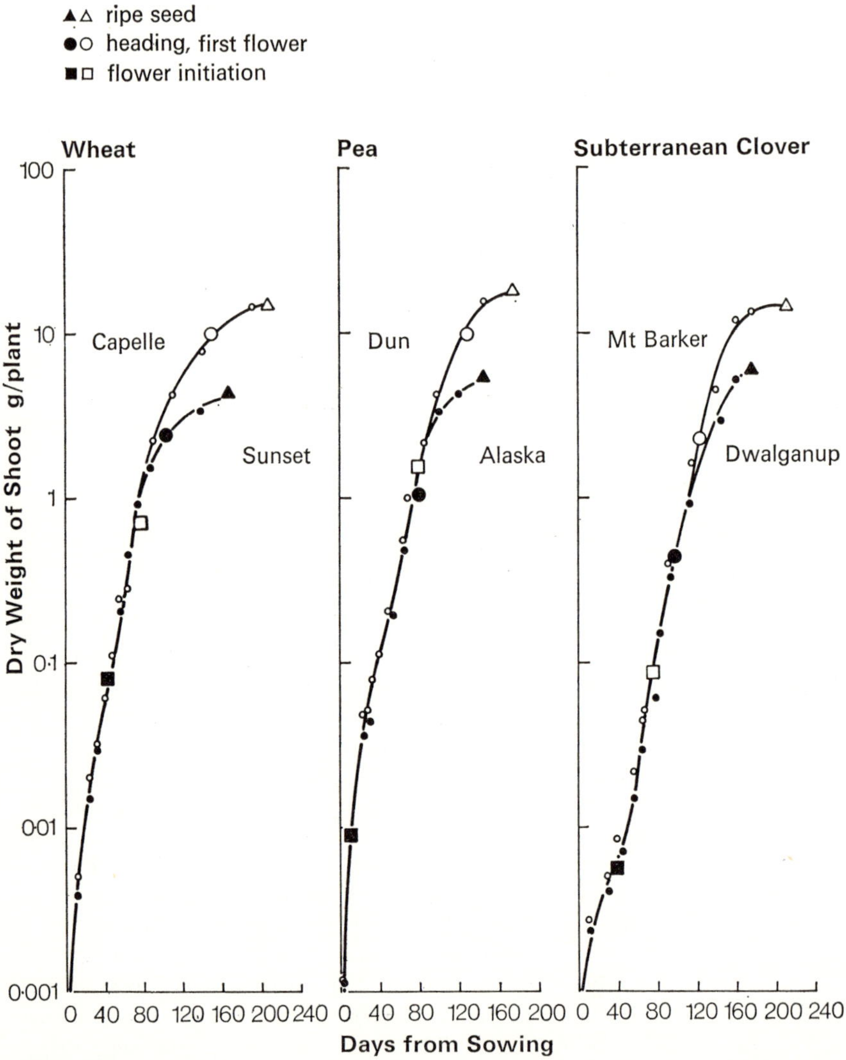

Fig. 3.2 In wheat, pea and subterranean clover the growth rate of the shoot begins to decrease at a time close to heading or first flower and hence sooner in the early variety (solid symbols) than in the late one (open symbols). Population: pea, 80 plants/m^2; wheat and subterranean clover, 200 plants/m^2.

varieties of three annual species (wheat, pea, subterranean clover) grown at Melbourne in the winter-spring season typical of southern Victoria. In each species the growth curve of the shoot of the early variety begins to flatten at flowering while that of the later one continues at a faster rate a little longer. In the extreme situation of an environment that arrests development and prolongs the vegetative stage, a high growth rate can be expected to continue for as long as conditions favour growth. This reaction is shown by late-maturing varieties of subterranean clover and of perennial ryegrass when sown in spring in middle latitudes (Aitken 1955a, 1966b).

From another aspect the rate of development is more tolerant than the rate of growth to the supply of nutrients. For instance, in both pea and wheat, only a slight reduction of soil fertility below the optimum reduces the growth rate. The effect is the lessening of leaf size and number of branches on the main shoot. However, the nutrient status must fall quite low before the time of flowering is affected, as under competition. The crowded plant is not only smaller than other more widely spaced plants but the rate of leaf appearance is slower. Because of this the leaf stage at FI is lower, the VN is reduced and the time of first flower or heading is later than in the less starved plants, although the time of FI may be unaffected. More intense competition might affect this last measure also.

Early development may even proceed to some extent in the almost complete absence of growth in certain varieties retarded or inhibited by high temperature—that is, having a 'cold requirement'. For example, if Dwalganup and Tallarook subterranean clover or Petkus winter rye are grown at high temperature, the vegetative stage (measured from sowing) can be shortened if the seed is held for some time at low temperature after imbibition but before sowing (Aitken 1955a, Collins 1972, Gregory 1938). In these and similar varieties, low temperature during the active growth of the embryo before the seed is mature affects the seedling in the same way.

It should be noted that a cold treatment (vernalization) of the seed or seedling slows growth at the shoot apex and may therefore lower VN without having increased development rate. An effective vernalization shortens the period from the start of the treatment up to FI compared with the same period in plants grown in high temperature from imbibition.

2 Structure of Shoot

The plant expresses its potential for growth and development through the structure of its shoot which with age may become more complicated by laterals arising from axillary buds. In early-flowering cereals the main shoot may have several subordinate laterals. In a complex shoot system such as that of subterranean clover the rates of growth and development may be retarded in the dwarfed main shoot by competition with the dominant laterals (Collins 1970). Another fact to be considered is that the proportion of the shoot system involved with reproductive activity varies with the species. In annuals all shoots are likely to become reproductive almost simultaneously. In herbaceous perennials the percentage varies and in woody perennials the yearly crop of reproductive shoots is only a small percentage of the total weight.

In the following discussion the shoot system is considered in its simplest form as the main shoot without laterals.

PHYTOMER

The shoot produces photosynthetic tissue and develops towards the reproductive state while it grows by a series of units or 'phytomers' initiated at the shoot tip (Fig. 3.3). The young phytomer when first formed at the shoot apex is small (about 0.1 mm high in many herbaceous species) and is made up of the ridge of the youngest leaf primordium and the narrow zone of the young internode which is only a few cells thick (Plate 1A).

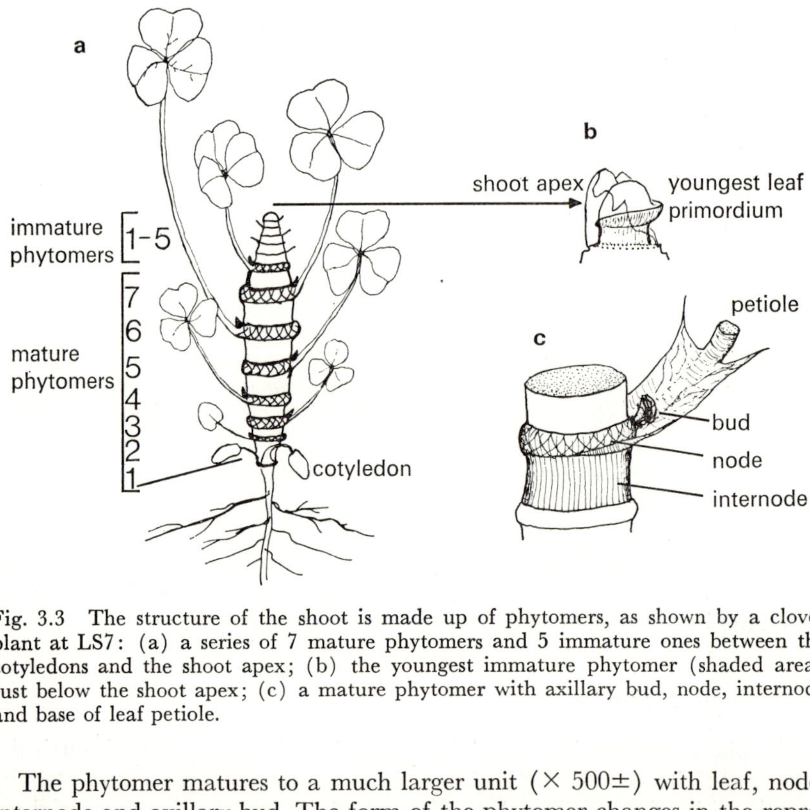

Fig. 3.3 The structure of the shoot is made up of phytomers, as shown by a clover plant at LS7: (a) a series of 7 mature phytomers and 5 immature ones between the cotyledons and the shoot apex; (b) the youngest immature phytomer (shaded area) just below the shoot apex; (c) a mature phytomer with axillary bud, node, internode and base of leaf petiole.

The phytomer matures to a much larger unit ($\times$ 500$\pm$) with leaf, node, internode and axillary bud. The form of the phytomer changes in the reproductive state with the degree to which leaf and axillary bud are modified into the reproductive structure, according to species. In pea and red and subterranean clovers, only the axillary bud of the reproductive phytomer becomes floral, as shown in Plate 2B for clover species. In subterranean clover the inflorescence begins as a simple bulge (marked 'a'). The bulge usually develops four floret primordia (one is marked 'b') and gradually forms the typical group of four florets on one stalk. The leaf primordium develops to the same type of leaf as on a vegetative phytomer. In red clover the one axillary inflorescence bulge forms many floret primordia (one is marked 'a') and becomes a terminal cluster of florets above its subtending leaf. In wheat and ryegrass the leaf primordium is also affected and only forms a small ridge below each spikelet at the reproductive node. Plate 1B shows that the wheat inflorescence begins as a series of axillary bulges, each one of which forms a spikelet of several florets.

Young shoot

An actively growing shoot is made up of both mature and immature phytomers (Fig. 3.3a). The phytomer becomes mature (active photosynthetically) when the leaf blade first opens. The mature phytomers are counted from the oldest (1) at the base of the shoot to the youngest with its just opened leaf (7), and the immature ones include the 5 increasingly younger ones up to the ridge of the youngest phytomer at the shoot apex. The figure shows a shoot made up of 12 phytomers (7 mature and 5 immature). This shoot can also be described as being at LS7 (leaf stage 7).

The final size of each phytomer depends on species, variety, nutrition, temperature and light. In addition the position of the phytomer in the series on the shoot is important. Each successive phytomer on the young shoot tends to be larger than the previous one. This can be seen in Figure 3.3a by comparing phytomers 1 and 5 for width of node, length of petiole and length of leaflet. The trend is reversed in the ageing shoot and is most pronounced in the shoot with axillary inflorescences as in pea (Fig. 3.10).

The time at which a particular part of the phytomer reaches maximum size depends mainly on the species and on temperature. In wheat, for instance, the length of the leaf blade on the youngest mature phytomer is finalized when the ligule can be seen at the base of the blade, but the sheath continues to lengthen for some days more. In clovers, as shown in Figure 3.3, both petiole and leaflets reach full size at about the time that the third-youngest leaf has just opened its leaflets. In other plants such as rape and beet the leaf continues to expand long after its first appearance as an assimilating green surface.

It follows from the above differences between species that observations on a character such as the rate of leaf appearance must be precisely defined (p. 19).

Reproductive shoot

When the shoot becomes reproductive, as shown by the bulges in the axils of the youngest leaf primordia (Plates 1B, 2), the activity of the shoot apex varies greatly according to the type of inflorescence. If terminal, as in the cereals and ryegrasses, the apex is absorbed by the developing spikelet on the uppermost reproductive node soon after the appearance of the 'double ridge' on all the reproductive phytomers. Figure 3.10 shows the structure of oat and ryegrass at flowering. Each plant has a definite number of vegetative and reproductive nodes and the topmost spikelet in both species can be seen. If the inflorescence is axillary, the apex continues to form reproductive phytomers much longer. The same figure shows that the structure of pea and subterranean clover towards the end of flowering includes a succession of reproductive nodes with leaves as well as pods or burs.

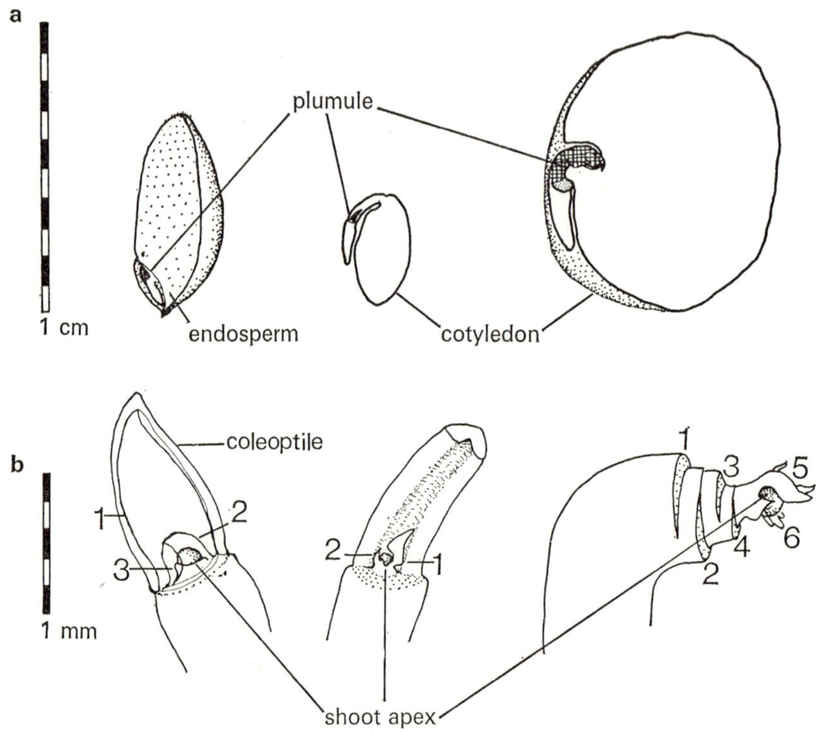

Fig. 3.4 The embryonic shoot (plumule) in the imbibed seed varies in size and structure, as shown in three species from left to right: wheat, subterranean clover, pea. (a) The plumule is shown in the seed as a shaded area of the embryo. (b) The dissected plumule reveals the shoot apex and the number of immature phytomers.

RELATIONSHIPS OF PHYTOMER INITIATION AND MATURATION

In the embryo of the seed the young shoot includes only the dome of the shoot apex and several very young phytomers. Figure 3.4b shows that wheat and subterranean clover have 3 and 2 phytomers respectively while pea has 6. The apex initiates 1 or 2 more between germination and emergence of the seedling above ground and also between emergence and the maturation of the first phytomer as indicated by the opening of its leaf, which marks LS1. For instance, perennial ryegrass and red clover both change from 2 in the embryo to 3 at emergence as shown by the number placed close to the youngest leaf primordium in Figure 3.5. These species produce a total of 5 phytomers (4 immature and 1 mature) by LS1. Maize changes from 6 in the embryo to 7 at emergence and to 8 (7 immature and 1 mature) by LS1. Pea is unusually rapid with 9 total phytomers at emergence of which 2 are already mature although inconspicuous because of their small scale-like size.

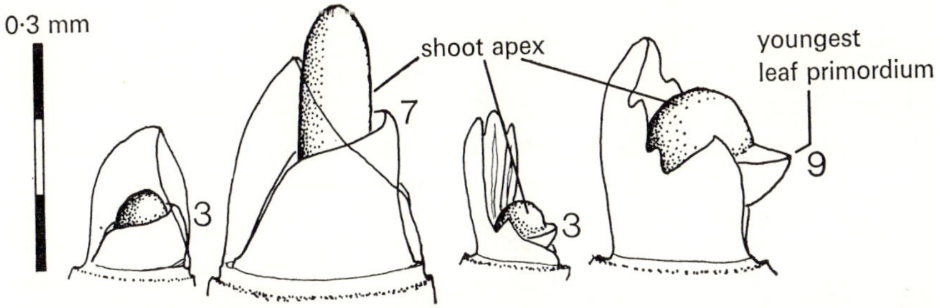

Fig. 3.5 When the plumules of seedlings emerge above ground, they differ in size of apex and in size and number of phytomers as shown from left to right in two grasses (perennial ryegrass and maize) and in two legumes (red clover and pea). The number of the youngest phytomer is shown to the right of the shoot apex.

Rates of initiation and maturation related to life of shoot

While the shoot grows actively, the rate at which the shoot apex initiates a phytomer is more or less closely related to the rate at which each one becomes mature, depending on species, leaf stage and stage of development. The rates are indicated by the inverse value of the number of days for each new primordium (plastochron) and for each open leaf (phyllochron). In general the rate of phytomer initiation increases a little faster than that of maturation with each new leaf stage. On the other hand, both rates are reduced, especially that of phytomer initiation, as the shoot loses vigour. This may occur because of low nutrient supply to the shoot tip, either because of competition from developing seeds towards the end of flowering or from laterals at an earlier stage.

The progress of the two rates in the shoot of early-flowering Alaska pea is shown in Figure 3.6, through the relationship of days from sowing to leaf stage and to total phytomers. The angle made by each sloping line with the base indicates the rate of formation of new units, a fast rate being shown by a wide angle and a slow one by a narrow angle. Taking the more obvious measure first, the slope for leaf stage (the lower line) reveals a fast rate of phytomer maturation as shown by rate of leaf appearance over the first 55 days after sowing. The rate then slows to a fairly abrupt stop because the immature phytomers above the top open leaf begin to die as the pods fill on the first-formed reproductive nodes on the shoot. The slope for total phytomers (the upper line) shows that new phytomers vary a little in their rate of formation as the shoot develops. The rate of phytomer initiation increases just after emergence of the shoot, then decreases after flowering begins at 35 days and ceases at 60 days although the apex remains alive for some weeks longer. The changing rate of phytomer initiation is also shown by comparing the number of immature phytomers on the shoot at intervals. The number is calculated from the vertical distance between the two slopes, and changes from 6 at germination to 8 at first flower and then decreases to 6 as the shoot apex slows in activity.

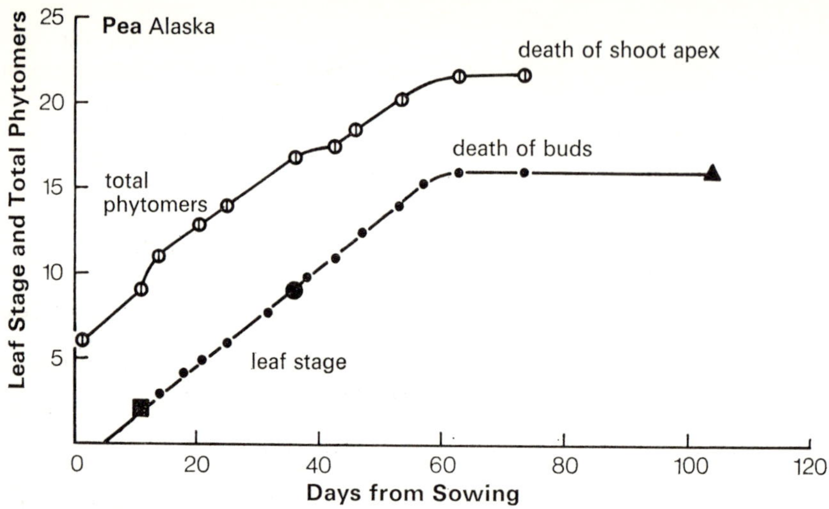

Fig. 3.6 In pea, both the leaf stage and the total phytomers on the shoot increase steadily with days from sowing until after the pods begin to develop, as shown in an early variety grown in summer in Alaska (site 10a), 1970. Developmental events: flower initiation (■), first flower (●), ripe seed (▲) on the early reproductive nodes.

Genetic variation in rate of initiation

Genetic variation may occur in the embryo and also throughout the vegetative and early reproductive stages or be more confined to one part of development such as the late vegetative stage.

General differences. Genetic variation in rate of phytomer initiation is well established in two plant families, Gramineae and Leguminosae (Aitken 1967), and may be expected in others.

Some species initiate leaf primordia faster than others (that is, have a faster plastochron), as shown both by the number of embryonic leaves in the seed and by the number of total phytomers in the older plant compared with the mature phytomers at any one time. Figure 3.7 gives examples from species within the grass and the legume family, grown under the same conditions and compared only in the vegetative stage. The number of immature phytomers in the embryo of the seed is indicated by the position of the symbols for each species on the base line. It can be seen that both maize and pea have 6 phytomers compared with only 2 in ryegrass and the two clovers.

The relationship of total phytomers and leaf stage (mature phytomers) is shown as a line for each species. In Figure 3.7a the more horizontal slope of the line for barley compared with that for ryegrass indicates a faster rate of leaf initiation. At LS1 in barley the total phytomers are 7 but at LS2 the total number is already 12. In ryegrass at LS1 the total phytomers are 5 and only increase to 7 at LS2. Thus twice as many immature phytomers are present in the barley seedling as in ryegrass at this stage. The slopes for wheat and maize indicate rates of phytomer initiation that are intermediate between those of

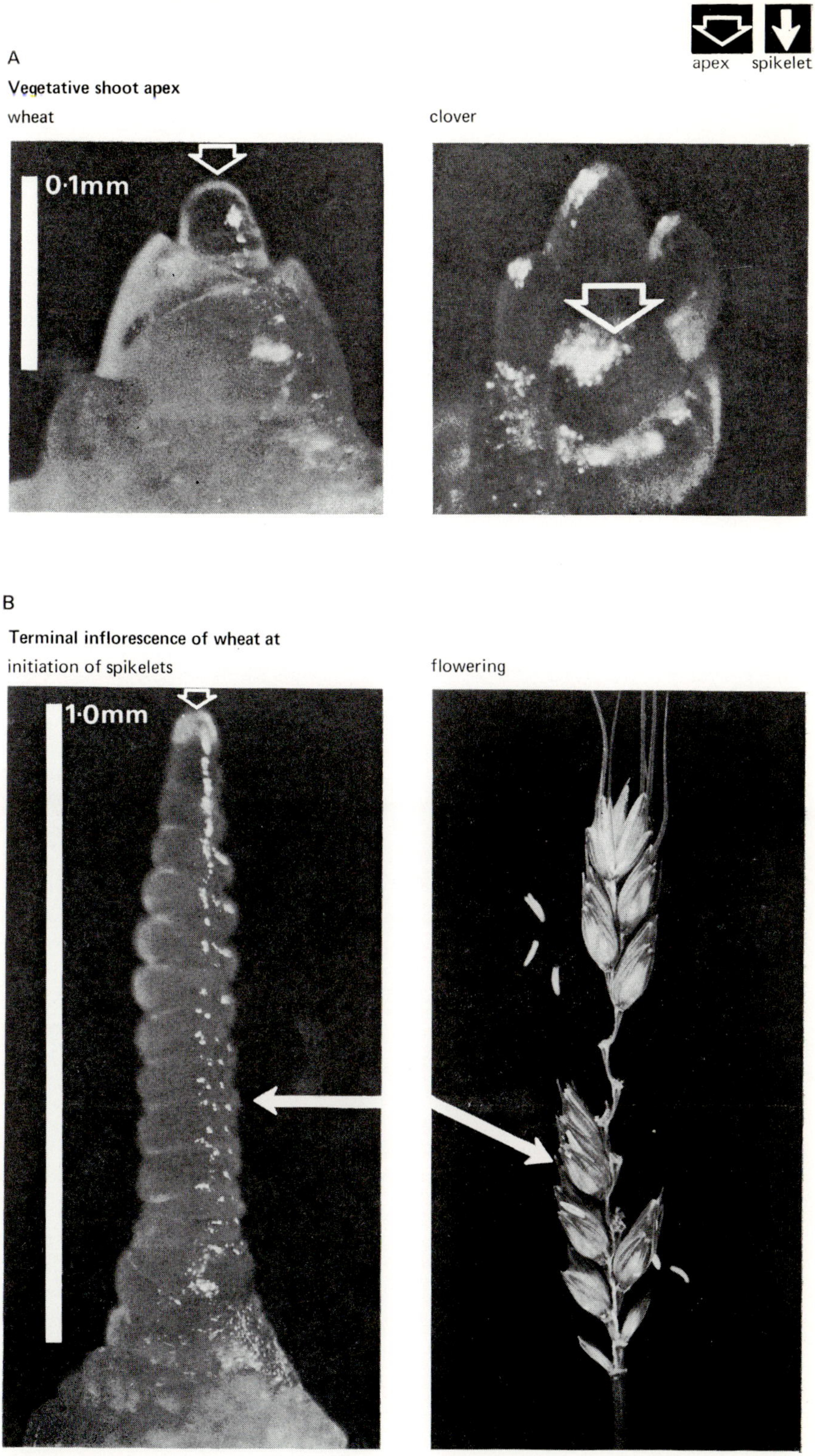

Plate 1 The appearance of the shoot apex in (A) vegetative cereals and clovers and (B) during reproductive development in wheat.

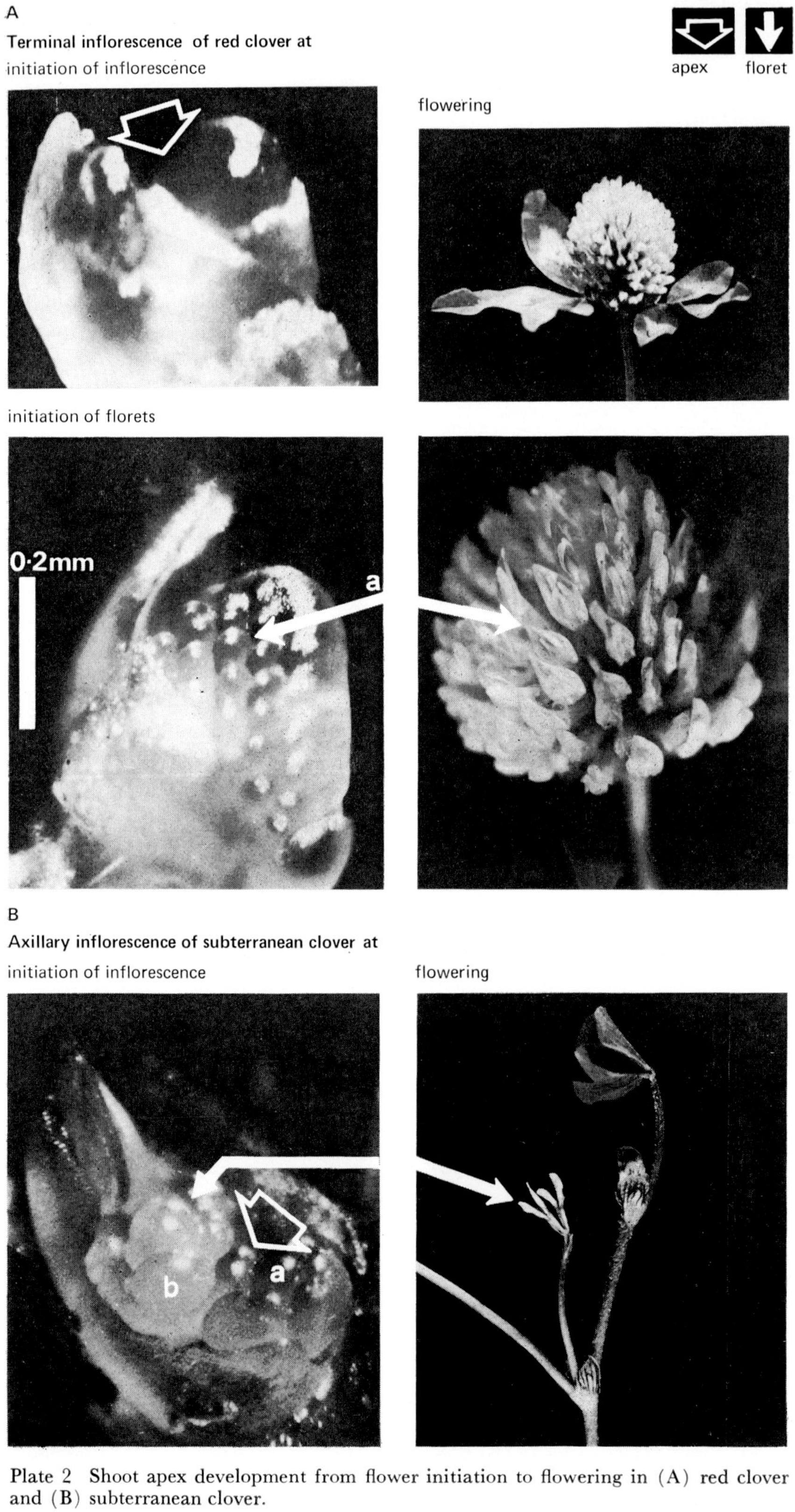

Plate 2 Shoot apex development from flower initiation to flowering in (A) red clover and (B) subterranean clover.

A

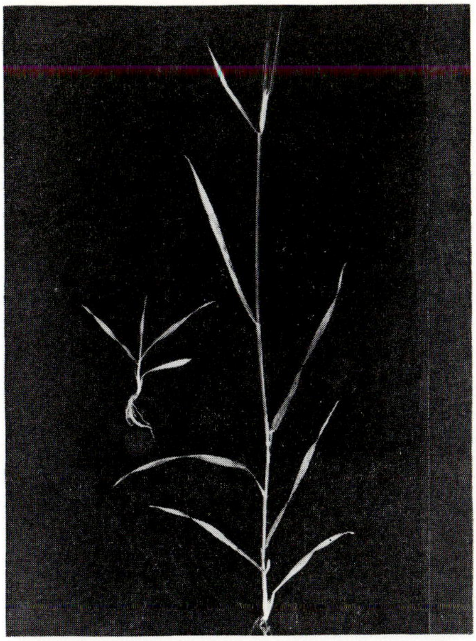

(a) (b)

The R1 response to temperature in Finlay barley after 52 days, sown summer 1970, Alaska, at (a) Pt Barrow 71°N (b) College 65°N.

B

(a) natural (b) 15 hr (c) 24 hr

The R6 response in Capelle wheat, sown winter, Melbourne. Treatments (a) natural (b) and (c) transferred after flower initiation to longer photoperiods and grown for 48 days. Arrow at leaf stage 11

Plate 3 Examples of plant responses which control the rate of reproductive development.

A

(a) (b) (c)

The R5 and R6 responses in Research barley lowered spikelet and grain number on the main shoot when grown under increased photoperiod after sowing early winter, Melbourne. (a) natural (b) 24 hr after flower initiation (c) 24 hr

B

The R3 response to temperature in Dwalganup subterranean clover, after 110 days, sown late autumn, Australia, at (a) Melbourne 38°S (b) Ayr 20°S

(a)

(b)

Plate 4 (A) Photoperiod responses in barley hasten flowering but reduce grain yield, and (B) a temperature-negative response in subterranean clover retards flowering but increases leaf yield.

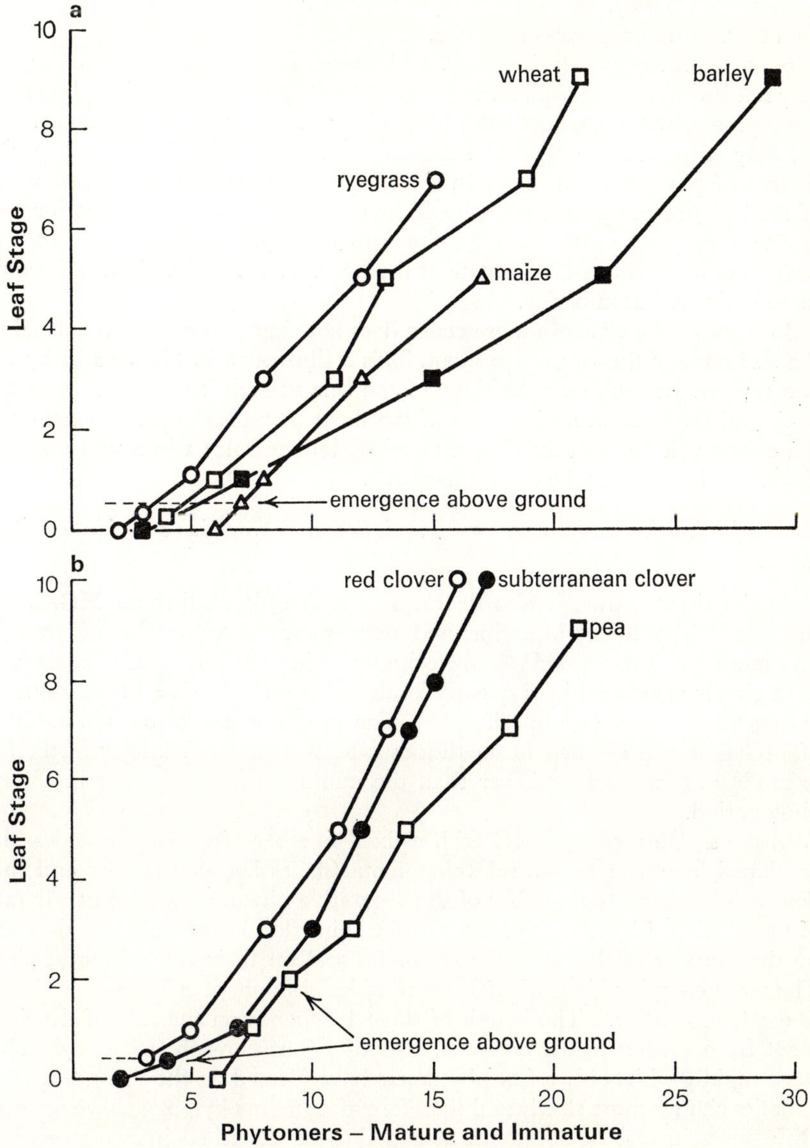

Fig. 3.7 The rates of phytomer formation and maturation differ with species as shown by the gradient of the line which relates total phytomers and leaf stage (equals mature phytomers) on main shoot in (a) four grasses and (b) three legumes. Each point represents the mean of ten plants with a standard error of 0·10 units. The values on the base line are derived from the seed.

ryegrass and barley. In Figure 3.7b the slopes indicate that the pea is faster than the two other species of legume.

Differences at late vegetative stage. The rate of phytomer initiation may or may not be affected in the late vegetative stage by the forthcoming change to the reproductive stage, according to species and to type of inflorescence.

Many grasses have a type of terminal inflorescence that involves a special pattern of phytomer initiation. In the early vegetative stage the relationship of total phytomers to leaf stage could be similar to that of the slope for ryegrass. With progress towards FI the rate of phytomer initiation becomes faster and more ridges accumulate at the apex. The slope would thus become more horizontal as in barley.

In legumes the type of inflorescence does not change the pattern normal to the leaf stage of the vegetative shoot. This is illustrated in Figure 3.7b by the two clovers. In each case the slope of the line scarcely changes between the early and the late vegetative stage of the shoot (from LS3 to LS10) although red clover will be soon forming a terminal inflorescence while subterranean clover will be forming the first axillary one.

Rates of initiation and maturation in relation to reproductive development of shoot.

As the shoot grows, it also develops at a rate depending on MG. Thus the rates of phytomer initiation and maturation, as a function of growth, are more or less related to development through time. The rates are most closely associated in very early kinds of MG which have a fixed number of vegetative nodes (see pp. 68, 111). The rates are least associated in kinds of MG that can develop in conditions otherwise not affecting growth, for example at temperatures lower than the minimum for growth or in a long photoperiod.

In the simplest case of MG with a fixed number of vegetative nodes, the combined length of the two developmental stages between sowing and first flower depends on the number of the vegetative phytomers and on their rate of maturation. Likewise the length of the third developmental stage depends on the number of the reproductive nodes and on their rate of maturation. The more complex kinds of MG are considered in chapter 5.

Stage 1, vegetative. The length of stage 1 depends on the rate of development from sowing to FI. Early varieties of a species reach the reproductive stage rapidly while only a few phytomers have formed on the shoot, but later varieties change more slowly and therefore after more phytomers have formed. Differences in rate of development can thus be described in two structural ways. The first measure is the *leaf stage* of the shoot when the first inflorescence initiates as a bulge in the leaf axil of the first reproductive phytomer at the shoot apex. The second measure is the *VN*, the number of vegetative phytomers at first flower or heading.

Figure 3.8 shows two early varieties, subterranean clover and wheat, at the beginning of FI with the first reproductive phytomers only visible by dissection of the apex. The structure of the apex of subterranean clover at this time is typical of many species of legumes with axillary inflorescences and

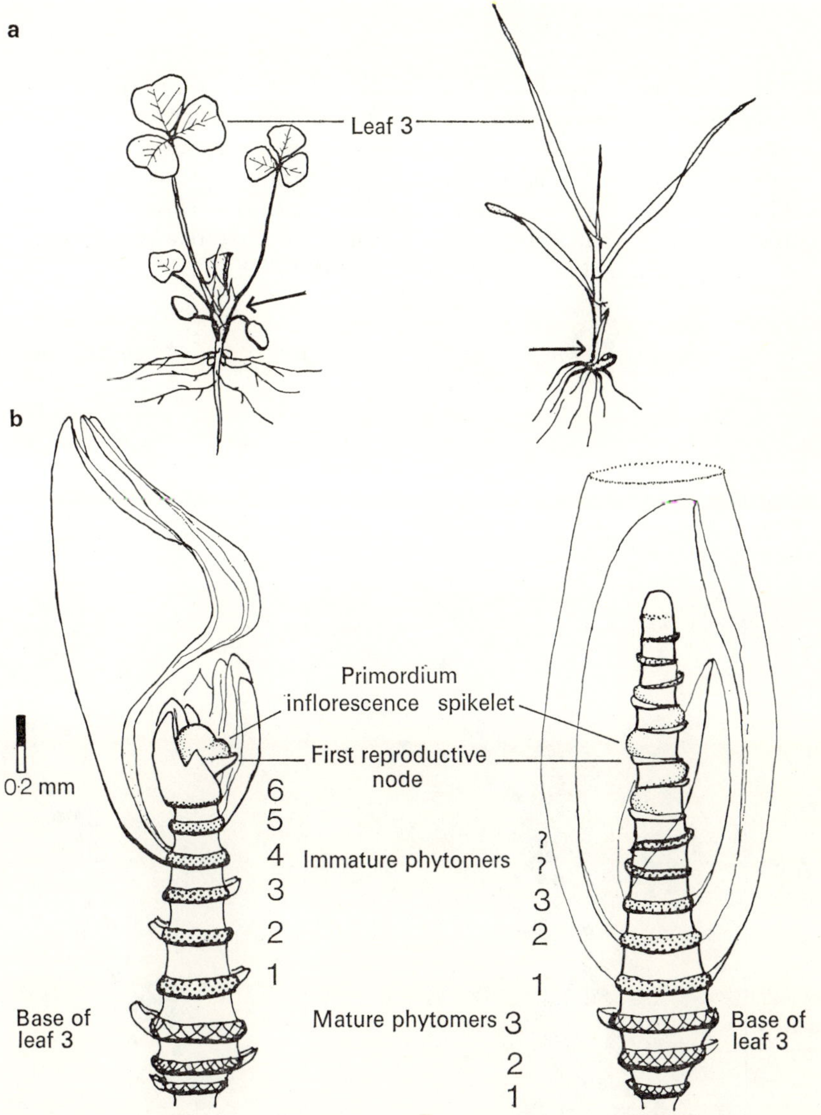

Fig. 3.8 The change to the early reproductive stage occurs at a leaf stage determined by the interaction of MG and environment and can be observed by dissection of the shoot tip. (a) Subterranean clover and wheat plants are compared at FI at LS3. The arrow at the base of each shoot marks the site of each diagram below. (b) Diagrams of the shoot in each plant show the number of immature phytomers between the top open leaf (the base of leaf 3 is shown) and the first reproductive node. The shoot is expanded vertically to show the nodes of the alternate leaves both mature (cross-hatched) and immature. Immature vegetative nodes are stippled and also those not yet reproductive.

also of others with terminal inflorescences like that of red clover. The structure of the apex of wheat is typical of grasses and of other plants with spike-like inflorescences.

The figure shows that both plants are at LS3 and also that the number of immature vegetative phytomers is 6 in the clover and more than 3 in the wheat. The VN of the clover shoot is therefore 9 (3 mature and 6 immature vegetative phytomers). The VN of the wheat is not morphologically obvious until the reproductive phytomers are further advanced. However, it may be estimated to be 8 (3 mature and 5 immature) from the relationship between leaf stage at FI and at heading as shown in Figure 3.9.

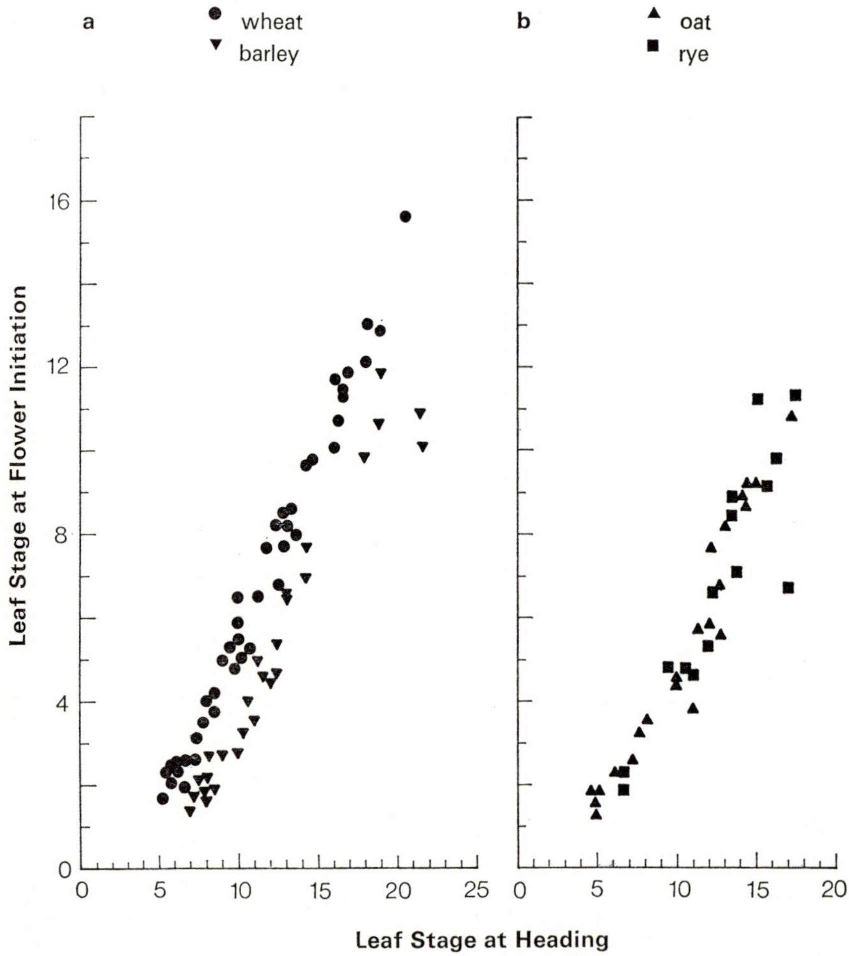

Fig. 3.9 The leaf stage on the shoot at FI in cereals determines the final leaf stage at heading in (a) wheat and barley, (b) oat and rye. The coefficient of variation for each value is about 1·5 per cent.

Stage 2, early reproductive. The length of stage 2 depends on the rate of development of the first reproductive phytomer from the appearance of the first inflorescence primordium to the mature inflorescence with ripe pollen and/or just fertilized ovules. The date of first open flower coincides approximately with the latter events in both out-pollinated legume species (red clover) and self-pollinated ones (subterranean clover, pea). In grasses, however, anthesis and fertilization usually occur a few days later than heading, the period being longer with lower temperature.

The rate of development of the reproductive structure is closely associated with that of the rest of the phytomer. The young immature phytomers between the top open leaf at FI and the first reproductive phytomer must also mature before the mature inflorescence appears. Hence the actual number of days for stage 2 will be a function of the general rate of maturation and of the number of immature phytomers up to the shoot apex at FI. These two facts can be used to estimate the length of the stage.

Factors affecting the rate of maturation are discussed in the next section. The number of young phytomers between shoot apex and top open leaf of the shoot is affected by species and by age and can be estimated by the relationships already discussed. In the case of subterranean clover at LS3 the total number of young phytomers is 7 (6 vegetative and 1 reproductive). In the case of wheat the number is about 5. With early varieties of pea the number is about 8. Other factors being the same, the larger the number of phytomers, the longer the early reproductive stage.

Stage 3, late reproductive. The length of stage 3 will depend both on the number of reproductive phytomers and on their rate of progress from fertilization of the flowers to mature seed. The type of inflorescence may therefore affect the length of the stage. Species such as wheat or ryegrass may take a short time because the spikelets at each reproductive node of the inflorescence flower and develop seeds at about the same time, whether the node is at the top or base of the spike. In species with axillary inflorescences the flowers on the last reproductive node may open many days after those on the first node, perhaps even when the seed at this node is almost mature, as in Collegian pea (Fig. 3.10) with its first reproductive node at LS13. There is therefore a tendency for species with axillary inflorescences to take longer in stage 3 than those with terminal inflorescences, where the growing season after flowering favours continued maturation of reproductive nodes.

FACTORS AFFECTING MATURE PHYTOMER FORMATION

The previous section drew attention to the number of phytomers on a shoot and their rate of maturation and discussed their importance to the length of the stages of development. In this section, factors that control the rate of phytomer maturation (as shown by the rate of leaf appearance) will be discussed.

Both environmental and genetic factors are important. Of the possible environmental factors, temperature is by far the most important one in the field when the crop is grown with adequate soil fertility, water supply, spacing of

Fig. 3.10 Because the shoot grows by a series of phytomers, the nodes can be numbered in sequence from the first vegetative node above the cotyledon (cot) or coleoptile (col) up to the first reproductive node. Laterals are numbered in the same way, with the first node counted as one, as shown on the lateral of subterranean clover, displayed above the main shoot. (a) Left to right: Collegian pea, Alpha oat, Wimmera ryegrass, Dwalganup subterranean clover. (b) The diagram shows the numbering system more clearly by expanding the closely packed nodes which are typical of the base of the shoot in the field. The species and varieties are in the same order as above.

plants, weed control and freedom from disease. Otherwise any one of these cultural factors, if it limits nutrients, water or light too much, could counteract the effect of favourable temperature.

Light, as related to photoperiod (hours of light per 24 hours), does not affect rate of leaf appearance. However, by influencing growth rate and photosynthesis the daily light intensity on the leaf surface may fall below the level for the optimum rate of leaf appearance at a given field temperature. This may occur because the mean daily radiation is too low or because of excessive shading in the crop.

The daily light intensity is almost half that of the daily solar radiation (the product of the mean hourly intensity of radiation and the number of hours of light) and is only likely to have a limiting effect at fairly low levels. The evidence from temperate species such as pea, subterranean clover, wheat and ryegrass suggests that the mean radiation of 150 $cal/cm^2/dy$* at Melbourne (38°S) in mid-winter is almost equivalent to that of mid-summer (600 $cal/cm^2/dy$) as far as rate of leaf appearance is concerned. However, the reduction of radiation to 40 $cal/cm^2/dy$ strongly retards the rate. Hence low light intensity must be considered as a limiting factor additional to that of low temperature. These conditions would apply to plants grown under the low radiation input of autumn and winter at high latitudes such as 52°N in Wales.

Temperature

The rate of leaf appearance is directly related to temperature, whether this factor is considered at steady levels under controlled conditions (Fig. 3.11) or in the fluctuating diurnal situation of the field (Fig. 5.3). The relationship is best observed in vegetative plants because the reproductive state in some species may affect the rate of leaf appearance in a more complicated way. The examples include not only four of the temperate species used in the climate studies but also one of the tropical species (maize) to represent plants with a much higher threshold for active growth.

Leaf stage in controlled temperatures. The relationships between temperature, days from sowing and leaf stage in the three species wheat, pea and maize are shown in Figure 3.11a-c. These relationships are considered from the aspect of the time taken to reach the same leaf stage. In (a), for example, each line gives for the same variety of wheat the relationship of temperature to the days from sowing needed to reach one of three leaf stages (1, 3 and 6). The three lines are alike and curvilinear because rate of leaf appearance is much more affected by temperature below 20° than between 20° and 30°C. The number of days from sowing to LS1 is about 20 at 10° and is reduced to 10 at 20° but only to about 8 at 30°. Thus the rate is doubled for the first rise of 10°, but scarcely changed for the second rise. The same relationship holds for the older plant at LS3 and LS6.

Other species also respond more rapidly to temperature in the lower part of the range. In pea (b) the curves include four leaf stages (3, 6, 9 and 12). Although pea is much faster in rate of leaf appearance than wheat, the

* 1 cal = 4·184 joules

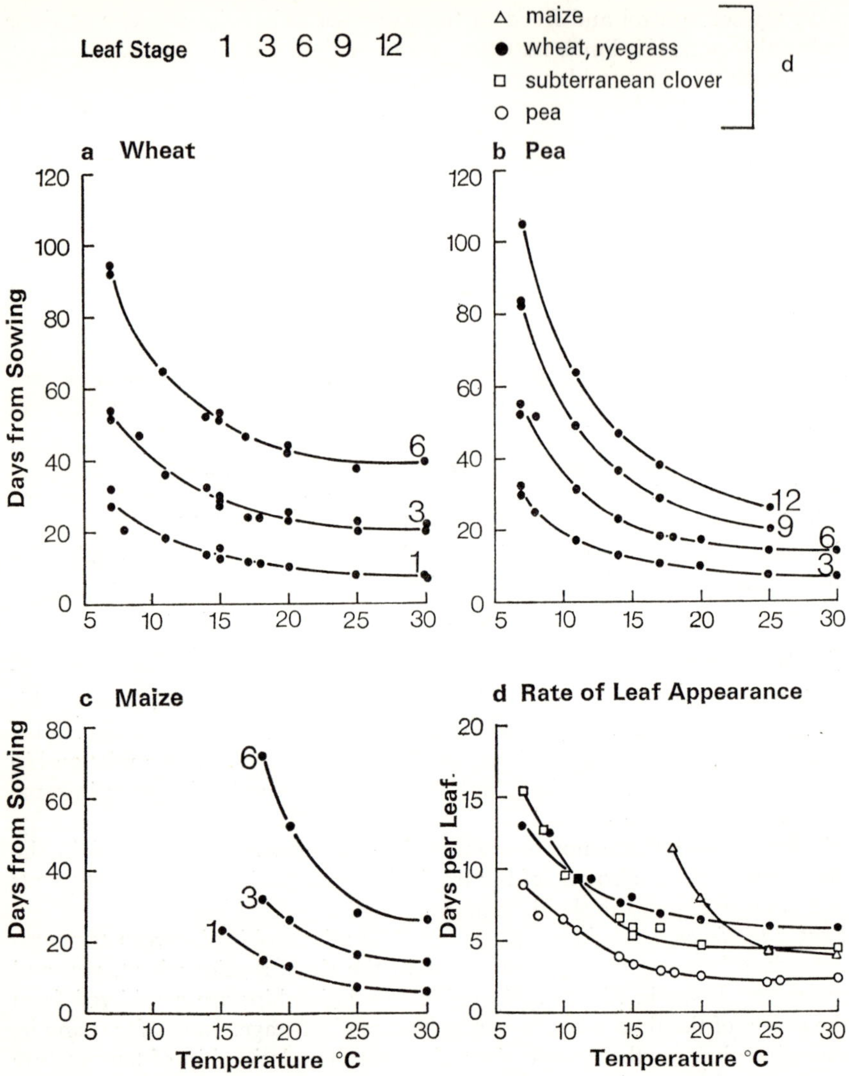

Fig. 3.11 The rate of leaf appearance on the shoot increases with higher temperature but the relationship differs with species. This is shown by the number of days from sowing to reach the same leaf stage in (a) wheat (b) pea and (c) maize, and by the number of days per leaf in (d) for maize, wheat, ryegrass, subterranean clover and pea. Each test was at constant temperature (± 1° per day) and a mean daily input of natural light of about 230 cal/cm².

relative response to temperature is similar. At 10° the period from sowing to LS3 is 20 days but only 10 days at 20° and slightly less at 30°. In (c) the curves for maize only begin at 15° but show the same trend of faster rate of change in the lower part of the span (15–25°) than in the higher portion (25–30°).

The importance of a plant's curve of response is emphasized in (d), where the mean number of days per leaf over the period from LS1 to LS6 is related to temperature. The most rapid increase of rate occurs between 18° and 25° for the tropical species, maize, and between 7° and 17° for the others, the temperate species.

Leaf stage in the field. In the field in southern Australia, except for the small area of high country, the important part of the plant's curve of response to temperature is that between 10° and 20°C. These represent mean values for months in winter and summer. One may expect that the time from sowing to the same leaf stage is halved at the higher temperature for temperate species such as wheat and pea. As the corollary, the same length of time in summer will allow the one variety to double the leaf stage reached from sowing compared with its progress in winter. For instance, in Figure 3.11, wheat reaches LS6 in about 45 days at 20° but only LS3 in the same time at 10°.

With tropical species their higher threshold for growth confines them to months with sufficiently high temperatures. In maize the time from sowing to LS6 is more than halved (from 75 days to 30) by the rise from 18° to 25°. Thus the rate of leaf appearance in this crop must be expected to be in the slower part of the plant's curve in southern areas of Australia because the mean monthly values in summer are usually very close to 20°.

Genotype

Genotype also affects the rate of leaf appearance but through differences between species rather than between varieties.

Species. Genetic differences between species may be large, as between pea and wheat. In Figure 3.11 (a and b) it can be seen that the slopes of the lines describing the relationship between time from sowing and leaf stage differ for the two species at any one temperature. The slope for pea is more vertical because the rate of leaf appearance is twice as fast. For instance, at 10°C and 40 days pea has reached LS6 while wheat has only reached LS3. Comparison of the slopes for pea and oat in Alaska (Fig. 3.12) shows a similar difference in rates. The late variety of pea reaches LS12 in 40 days but the late oat only reaches LS6 in the same period of time.

The contrast between wheat and maize (Fig. 3.11a, c) is even more striking. Because the latter is a tropical species, the comparison can only be made between 15° and 30°. At 15° maize is not only very slow but unhealthy. At 18° it is healthy but still slower than wheat to reach LS6 (70 days compared with 45). The position is reversed at 25°, for maize reaches the same stage in less than 30 days but wheat takes 40 days.

In other species the rate of leaf appearance may differ little or not at all between species over the same range of temperature as shown by the common curve for the days per leaf (Fig. 3.11d) in wheat and perennial ryegrass.

Variety. Surveys within a species such as pea or subterranean clover (Aitken 1945, 1955a) or perennial ryegrass (Cooper 1956a) have shown a widespread uniformity in rate of leaf appearance. The few varieties that are a little faster have a smaller leaf. This linkage reduces the value of the faster rate of

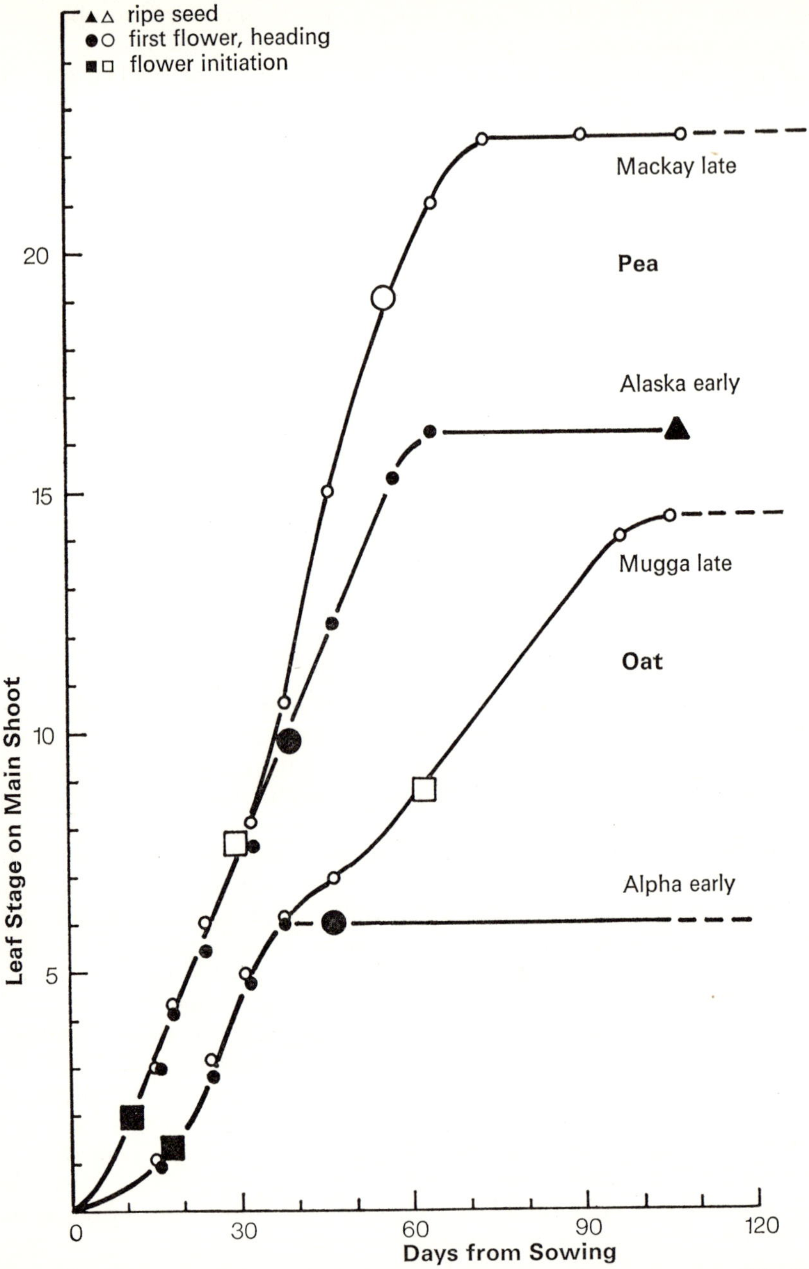

Fig. 3.12 The leaf stage on the main shoot increases at the same rate with days from sowing in both early and late varieties of the one species, until the early variety approaches the late reproductive stage. This is shown in pea and in oat, grown at College, Alaska, in summer 1970. In pea the flowering nodes of the early variety mature leaves more slowly than the less advanced late one. In oat the early variety reaches its final leaf stage rapidly, and before the late one becomes reproductive. The latter produces leaves more slowly until killed by frost (dashed line).

leaf appearance to the daily leaf area and growth rate as shown for perennial ryegrass (Edwards 1963).

Since the reproductive stage may complicate the rate of leaf appearance in otherwise uniform varieties, this character is best observed over the period from sowing to FI or part way towards flowering, after which the earlier-flowering varieties become slower. This is shown in Figure 3.12 with early and late varieties of two species grown in summer in Alaska. In pea the two varieties are similar in leaf stage until the early one begins to flower, while in oat the two varieties are alike until the early one produces its final leaf before heading.

3 Structural Characters of Shoot, and Growth and Development Rates

Whether concerned more directly with growth or with development of an organism, the agronomist is interested in the yield of useful plant material. But from the aspect of climate as discussed in this book, development must be mostly confined to the rate of progress of the plant towards maturity.

It is nonetheless instructive to consider the measures used in the 'growth analysis' of crop yield (Watson 1968) and those used in the 'developmental analysis' of the growing period of a crop. Table 3.1 compares the components

Table 3.1
A comparison of components and characters affecting growth and development rates (columns relate vertically only)

Measures	Components	Characters
*Growth rate per plant**		
relative g/g/dy absolute g/dy	net assimilation rate: mg/leaf cm^2/dy leaf area ratio: leaf cm^2/plant g	leaf number and size number of laterals rate of leaf appearance
Developmental rate of main shoot or vigorous lateral		
per stage: inverse value for days/stage per growing period: inverse value for days/growing period	inverse value for VN rate of leaf appearance rate of seed maturation	minimum VN per MG extra VN from developmental responses rate of leaf appearance and flower development rate of seed maturation

* Blackman 1968

of growth and of development and lists the plant characters that contribute to both measures.

The components of growth rate provide a daily amount of photosynthetic product through the characters affecting the quantity of active leaf area. In contrast the components of development rate control the number of days taken for each developmental stage. A low VN, a fast rate of leaf appearance and a fast rate of seed maturation result in a short growing period. It should be noted that rate of leaf appearance is important to both types of analyses for different reasons. A fast rate increases the number of new leaves within a given time and thus may affect the area of active new green tissue and the growth rate. At the same time a fast rate of leaf appearance shortens the period from sowing to flowering in early varieties with a fixed number of vegetative nodes.

COMPONENTS OF DEVELOPMENT

Table 3.2 shows the effect of increasing each developmental component on the length of a developmental stage. The faster rate of leaf appearance shortens the two stages between sowing and flowering while the faster rate of seed ripening shortens the third stage.

Table 3.2
Components of development in relation to length of stage

Component increased	Length of development stage		
	Stage 1	Stage 2	Stage 3
Vegetative nodes on shoot (VN)	increased	unaffected*	unaffected
Rate of leaf appearance	decreased	decreased	unaffected
Rate of seed ripening	unaffected	unaffected	decreased

* varieties of a species with similar rates of phytomer initiation and maturation

The VN may be a fixed value in the one variety as in those which reach FI very early, such as Alaska pea and Early Wonder oat. In the more usual case, VN varies with the phenotype, because of variation in developmental rate. A higher VN indicates delayed FI and hence a longer first stage, given no contrary interactions. The second stage could also be slightly increased if the number of immature phytomers above the leaf stage at FI were raised at higher leaf stages. In the pea (Fig. 3.7) an early variety which reaches FI at LS2 has already 9 phytomers and 7 of these must mature before first flower, while a later variety which begins to initiate at LS8 has formed the twentieth phytomer and must mature 12 phytomers before first flower.

IV Climate and Growing Season

This chapter examines the concepts of climate and growing season in relation to plant development and describes the types of growing season used in the author's climatic studies.

1 Concepts of Climate and Growing Season

CLIMATE

The climate of any one place is made up of the daily and the seasonal weather patterns from year to year. These patterns result from daily changes in solar radiation and from the atmospheric systems controlling cloud and rain.

Climate is described from meteorological measures some of which have been taken over a long period of time in any one place. The more common measures include hours of sunshine, temperature, rainfall, humidity, and wind velocity. Measures of solar radiation (global, diffuse and net) and evaporation are harder to obtain. The number of hours of light per day is also important but can be predicted if the latitude is known.

Different places with similar climates can be generalized into a climatic zone based on temperature and rainfall. These zones cover the earth from the tropics to the poles. Within each zone the climates must be subdivided and grouped according to the cooling effect of altitude and the amount and distribution of rainfall through the year.

GROWING SEASON

From the point of view of the plant's survival and yield from year to year, the growing season is the most important aspect of climate. The concept of growing season includes three features: soil water supply, temperature and the length of time over which these apply. Given adequate soil water during the season, the temperature is very important, firstly because of its general effect on growth rate (Davidson 1969) and secondly because of its interaction with plant type.

The growing season is the number of consecutive months during the year

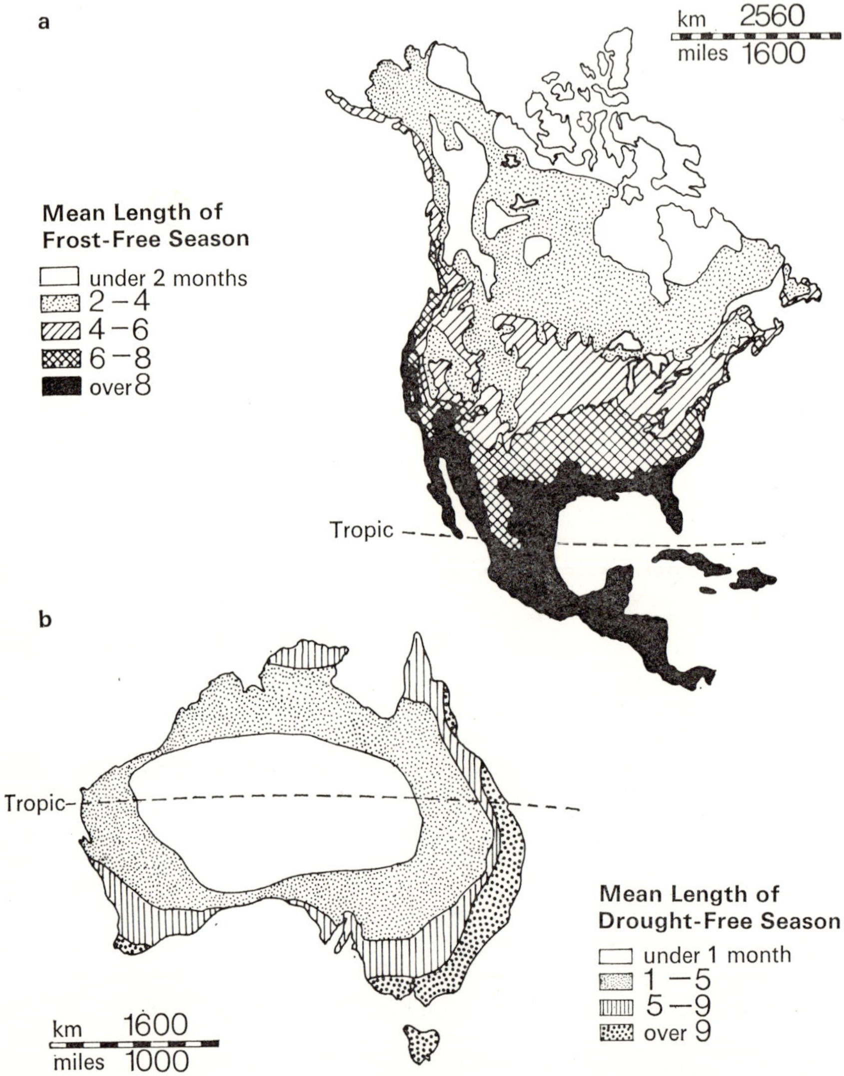

Fig. 4.1 The concept of growing season varies with geographical region. Frost in North America or drought in Australia is the limiting factor to the length of the season. (a) The isolines define zones differing in the number of frost-free months per year. (b) The isolines define zones differing in the number of drought-free months per year (the consecutive months of effective rainfall).

The map of America is adapted from the *World Atlas of Agriculture*, vol. 3 (Istituto Geografico de Agostini, Novara, Italy, 1969).

that are available for active plant growth because of sufficient water and favourable temperature. However, the threshold temperature for growth depends on the type of plant, tropical species having a higher threshold than temperate ones. Hence a growing season of a given length could be entirely suitable for a temperate plant though only a portion of it might have high temperatures and not be long enough for a tropical plant.

Length of growing season

The growing season includes the whole year in favourable climates such as the coastal areas of south-eastern Australia. In other places it may be less than the minimum length needed for the growing period of even the most rapidly developing crops, as in inland southern Australia with less than a five-month growing season (Fig. 4.1b). The main limiting factor to the length may be either water or temperature, or both, depending on the climatic zone.

Water as the limiting factor. Shortage of soil water tends to be the major limiting factor in the tropics and the warm temperate regions because the high level of evaporation makes a given quantity of rainfall less effective than in cooler regions. The rate of transpiration in plants is closely related to that of evaporation. Therefore a relationship between monthly rainfall and evaporation is used to define the months of effective rainfall for plant growth (Leeper 1970).

Figure 4.2 gives two examples of a growing season in Australia as indicated by the consecutive months of effective rainfall superimposed on the curve of the mean monthly temperature. The season at Ayr in the tropics is six months long, beginning in summer with the first rains and continuing until the drought begins in winter. The monthly temperature is high (>20°C) through most of the year. In contrast the season at Melbourne in the temperate zone is longer but cooler. It begins in autumn with the first rains after a more or less dry summer and continues for about nine months until the end of the following spring. The temperature is low almost all the time compared with the tropical example, changing from 14° in the autumn to about 9° in mid-winter and to 17° in the early summer.

Temperature as the limiting factor. Low temperature is the major limiting factor to the length of the growing season in the cool temperate and the polar regions. The season is defined as the number of frost-free days, and the number in North America (Fig. 4.1a) ranges from more than 240 days in the warmer south to less than 60 in the cooler northern states. The number of frost-free days provides a growing season more or less suited to the temperate crop species according to the number of months above a threshold such as a mean monthly temperature of 5°. In 1970 at College in central Alaska, for instance, the growing season was limited to about 120 days above 5° between the last frost in early May and the first in early September. In this part of Alaska the growing season was limited even further by several weeks of drought after the last frost in spring. The growing season of 120 days, judged on temperature above 5°, was thus shortened to about 100 days for crops that had to wait for the rains before germination could begin.

The efficiency of the temperature of the growing season must be better assessed if increasing the growth rate of crops is the aim (Newman 1959), though the present study on plant development does not justify this more complex approach.

Quality of growing season

Given adequate water during the growing season, other features such as temperature must be considered, depending on whether the investigator is concerned primarily with growth rate or development rate of the plant.

Factors affecting growth rate. The growth rate (dry weight/plant/day) is related to the daily input of energy and to associated factors of temperature, light intensity, light quality and day length through the photosynthetic activities of the plant. Growth rate increases positively with energy input, temperature and light intensity between lower and upper boundaries, the level of which depends strongly on genotype.

The positive relationship of growth rate to temperature is probably the most important one in the wide range of growing seasons used for agricultural production. Genotype affects it through controlling (a) the temperature levels for the start and finish of active growth and (b) the relationship of temperature during active growth to the growth components, which are net assimilation rate and leaf area ratio.

Plants are conveniently ranked as temperate or tropical types, according to the lower boundary for active growth which is about 5°C for temperate plants and about 15° for tropical ones. The optimum temperature is likewise higher for the tropical plants and also the upper boundary for active growth. In addition this group is much more frost-sensitive.

Thus a farm in the temperate zone could have two types of growing season if sufficient water were available. At Melbourne, for instance, the first season would begin in winter at 10°C and the second one in spring at 13° (Fig. 4.2). A temperate crop such as wheat or pea could grow in both types of season but at a higher rate at the higher temperature. A tropical one such as maize could not grow at all if sown in winter and would grow very slowly at first when sown in spring.

In high latitudes such as in Wales 52°N, a farm near the coast would have a moderately warm spring-summer growing season. The mean monthly temperature would be above 10° for the six months of the spring and summer but with no month above 15°. Such a growing season could be below the optimum for maximum growth rate even for temperate species.

Factors affecting development rate. Development rate is inversely related to the length of the growing period (Table 3.1). Both temperature and photoperiod may be important and may react with the plant's potential development rate in a complex way. Development rate measures the rapidity of change from the beginning of the vegetative stage to the end of the two reproductive stages and not the relationship between the quantity of energy used by the plant and the quantity of reproductive tissue produced. Yet development can only proceed if there is a certain minimum of healthy growth from adequate photosynthetic activity.

Development rate depends partly on the processes of growth which affect

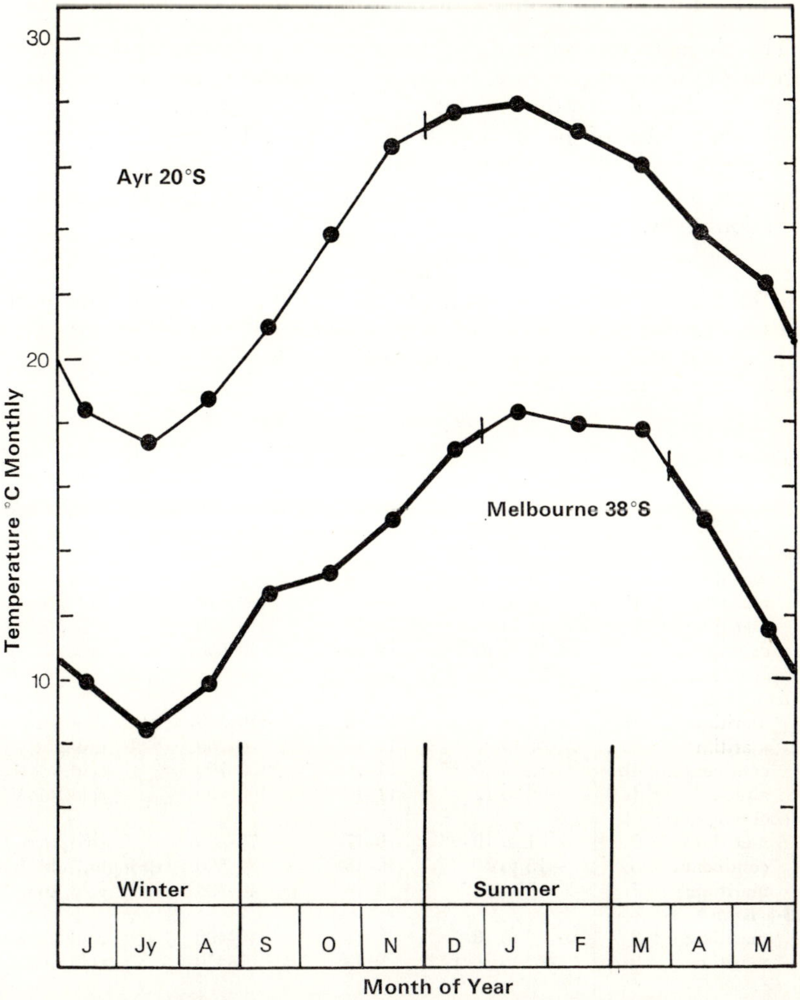

Fig. 4.2 In Australia the growing season may vary greatly in temperature according to the latitude of the locality and the time of year. This is shown by the hot wet season at Ayr (site 3a) in the tropical farming zone compared with the cool wet season in the temperate zone at Melbourne (site 3b). Each circle shows the mean daily temperature per month and the thick line indicates the growing season (the period with adequate soil moisture).

the rates of leaf appearance, flower and seed formation through response to temperature and partly on the processes of development which direct the plant towards flowering, fertilization and seed formation. These internal controls give an intrinsic tendency to flower and also may cause the plant to respond to temperature and photoperiod in a special way. These factors act more as signals for hormonal changes which hasten or retard the rate of development rather than as direct controls.

2 Measurement of Climatic Factors Affecting Development Rate

The three factors selected for measurement at each experimental site (Table 2.2) were temperature, photoperiod and solar radiation. The range of monthly values during the year for each factor is shown in Table 4.1 against the index number of the site and its climatic type. The mean daily temperature per month is used in this table and throughout because it is a reasonable approximation of the daily temperature at plant level, as the following observations show.

Table 4.1
The location of the experimental site affects the range through the year of the mean monthly values of climatic factors and the limiting factors to plant growth

Latitude zone*	Site	Yearly range climatic factors mean daily/month‡ Temperature °C	Photoperiod hr/dy	Radiation cal/cm²/dy	Limiting factor to growth
North tropical					
maritime	1a	23 to 26	12–14	360–550	drought
maritime	1b	17 to 18	12–14	390–610	drought
maritime	1c	11 to 15	12–14	410–670	drought
continental	2	10 to 18	12–14	440–550	drought
Middle latitudes					
temperature					
maritime	3b	9 to 19	11–16	170–580	drought
maritime	4a	7 to 20	11–16	110–760	drought
continental	4b	2 to 20	11–16	100–700	cold, snow
alpine	4c	−6 to 12	11–16	100–750	cold, snow
cool temperate					
maritime	5	1 to 19	10–17	70–580	cold, snow
continental	6	−15 to 20	9–18	80–590	drought, cold, snow
maritime	7	3 to 15	9–18	40–390	cold, snow
Sub-antarctic†					
maritime	8	3 to 8	9–19	40–400	cold, snow
maritime	9	−2 to 6	9–19	?–520	cold, snow
Arctic					
continental	10a	−24 to 16	7–24	7–520	cold, snow, drought
tundra	10b	−28 to 3	3–24	0–520	cold, snow, drought

* Strahler 1970, adapted
† Holdgate 1970
‡ value to nearest unit

SOLAR RADIATION

Solar radiation (global—that is, from sun and sky) was available from meteorological stations close to most of the sites or else at sufficiently similar places to be used for estimated values (Appendix 3).

The daily global radiation on the plant supplies both light energy for photosynthesis and heat energy which affects the plant via the temperature of the tissues and the surrounding air and ground.

The daily radiation is not directly important to plant development except at marginal levels because of the small percentage of energy used for development and because of the heat losses from air and ground that affect the temperature of the plant. However, an outline of the information from the experimental sites provides as essential background to the understanding of temperature and photoperiod in relation to development.

Radiation on clear days

The maximum possible radiant energy per day at any place varies with latitude through the factors of intensity per hour and the number of hours of light per day. The intensity is lowered with distance from the equator but the role of the daylight hours depends on the latitude and the season. The number of hours scarcely varies throughout the year in the tropics. At higher latitudes the decreasing number of hours in the winter reduces the daily input of radiation while the increasing number of hours in the summer compensates to some degree for the lower intensity. Therefore, with increasing latitude, (a) places compared at the same season have lower maximum intensity, and in winter total radiation is even lower because of the fewer hours of light and (b) at the one place the contrast between seasons at high latitude is intensified by the few hours of light in winter.

Figure 4.3 gives examples of the above trends by comparing clear (or almost clear) days in the same season at site 2 (Chapingo 20°N), site 3b (Melbourne 38°S) and site 10a (College 65°N). A set of curves at each place gives the hourly radiation for particular days of the year.

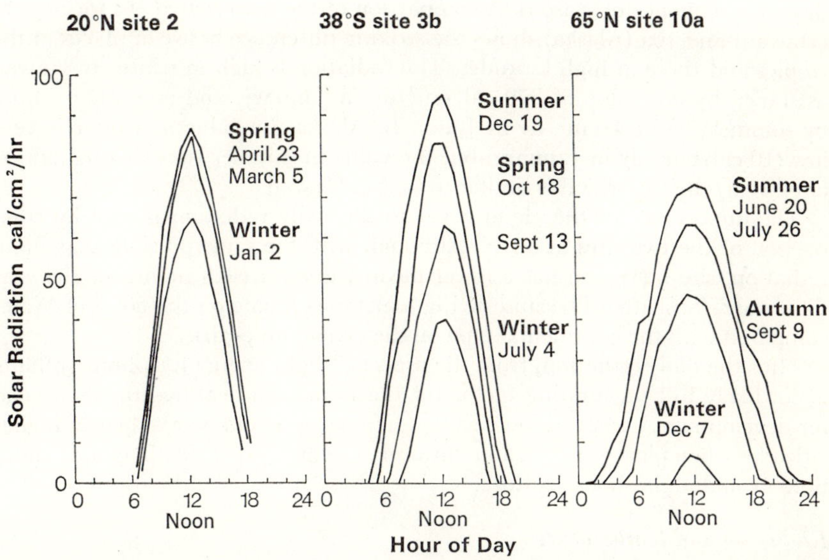

Fig. 4.3 The daily input of energy from the sun at plant level on clear days varies greatly throughout the day and between the seasons, according to latitude. The selected days at site 2 (Mexico), site 3b (Melbourne) and site 10a (Alaska) show the curve of the hourly solar (global) radiation from sunrise to sunset and the increasing difference between seasons, with higher latitude.

Winter. The daily curves for winter (the innermost curve of each set) differed much more between the places than those for the other seasons. Inspection of the curves for 2 January, 4 July and 7 December shows that the area within each curve is greatly reduced with increased latitude, both because of lowered intensity per hour and fewer hours of light. Thus the maximum intensity per hour at Chapingo was 65 cal/cm^2/hr compared with 7 at College and the hours of light were about 12 compared with 4.
Spring and autumn. The differences in daily radiation between the three places are less in spring and autumn. A comparison of the daily curves on 5 March, 13 September (spring) and 9 September (early autumn) shows that the maximum intensity falls from 85 cal/cm^2/hr at Chapingo down to 45 cal/cm^2/hr at College, but the hours of light are about the same. Hence the total radiation is only affected by the difference in intensity.
Summer. The differences in daily radiation between the three places are lessened further in summer because of the compensating effect of the longer day length. The comparison between Melbourne on 19 December and College on 20 June shows a lower maximum intensity (75 compared with 95) at the higher latitude but also an extra 5 hours of light.

Daily radiation through the year

The actual mean daily radiation per month through the year at the experimental sites is shown in Appendix 3. The mean daily radiation for a month is usually less than that for a clear day because of the frequency of cloudy days. Nevertheless the same trends occur as shown in the previous section for clear days in a sequence of seasons. A comparison of the monthly values for sites 1a (Hawaii) and 10a (Alaska) shows the striking difference between places in the tropics and those at high latitudes. The radiation is high in winter in Hawaii as shown by the value of 370 cal/cm^2/dy in January, and rises only a little by summer (550 cal/cm^2/dy in June). In Alaska the value in winter is very low (10 cal/cm^2/dy in January) but the value in summer rises to a maximum of 520 cal/cm^2/dy which is similar to that in Hawaii.

The importance of the cloud cover to the daily radiation is seen by comparison of the two sites at 52°N. Although with the same possible maximum radiation, site 7 (Wales) has a lower mean value for each month of the year than site 6 (Saskatoon) because of the maritime climate on the coast of Wales compared with the continental type on the Canadian prairie.

Thus the daily radiation, from the aspect of light energy for photosynthate, is not likely to be a limiting factor for the minimum healthy growth needed for development except at low levels. These levels could be reached in higher latitudes when plants are sown in autumn or winter, especially in cloudy situations or under glasshouse culture.

Radiation and temperature

In temperate agriculture, radiation and temperature are closely associated in the seasonal changes during the year. This association is weak or absent in other types of growing season. The experimental sites were therefore located to include not only the growing seasons commonly used for food production

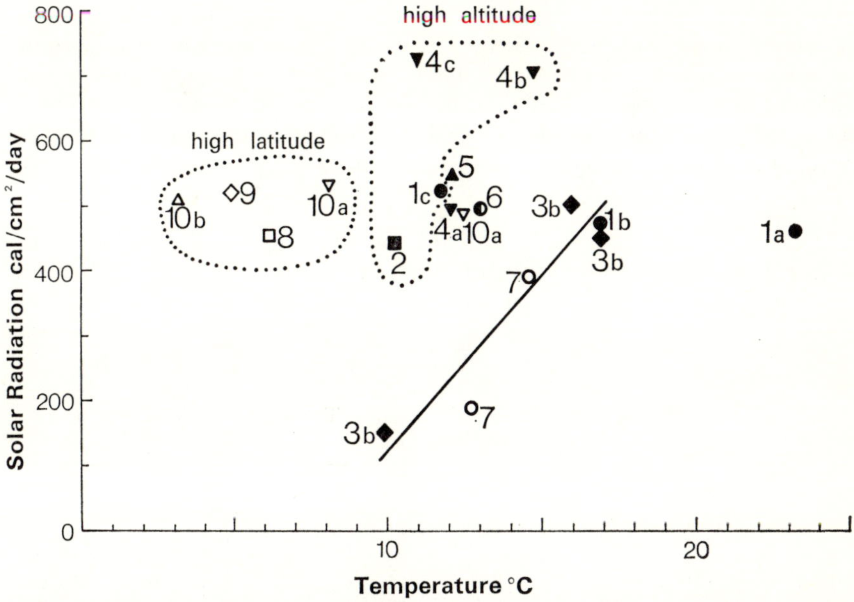

Fig. 4.4 The mean daily values of temperature and solar (global) radiation are shown for the first month after sowing at all sites. Low radiation and low temperature are commonly associated in agricultural regions in the temperate zone, as found at site 3b in the winter sowing at Melbourne 38°S and at site 7 in the autumn sowing at Aberystwyth 52°N. The association is broken at high latitudes and altitudes, as illustrated by the encircled sites.

in the temperate zone but also tropical ones and those marginal for plant growth at high altitudes and latitudes.

In Figure 4.4 the mean daily temperature and radiation for the first month after sowing is shown as a symbol for each site. The line shows the trend of temperate growing seasons at sites 3b and 7, from low values of temperature and radiation in winter and autumn respectively to higher values in spring and summer.

In other places, the seasonal association is reversed or missing. On sub-antarctic Macquarie Island (site 8), radiation is high during the summer growing season but the mean temperature of the first month is only 6°. Radiation is also high during the winter on the tropical island of Oahu in Hawaii (site 1a) but the mean temperature is 23°. Other comparisons may also be made, as with the cool growing seasons at Melbourne (site 3b) and at Timberline in California (site 4c). The temperature is close to 11° at Melbourne after sowing in winter and at Timberline in late summer but the radiation differs greatly (150 and 720 cal/cm^2/dy, respectively).

Radiation and development

The climatic studies suggest that radiation is not an important factor in plant development in the more common growing seasons. For example, the growing period of early varieties of pea and cereals is more closely related to mean daily temperature (Table 10.1) than to radiation. However, the more extreme growing seasons provide evidence that radiation may be important in situations where it raises the temperature of the shoot apices close to the ground several degrees higher than the air at standard screen height. These situations occur in growing seasons with high radiation but moderate to high heat loss as in places at high latitude and altitude.

Table 4.2
The effect of temperature and radiation on the length of the periods from sowing to LS3 and to flowering in five varieties of barley and in Alaska pea and Early Wonder oat (coefficient of variation less than 1 per cent)

Site	Time sown	Sowing to LS3			Sowing to flowering			
		Temperature °C/dy	Radiation cal/cm²/dy	Barley no. of days	Temperature °C/dy	Radiation cal/cm²/dy	Pea no. of days	Oat no. of days
10b	summer	2·5†	380	50	—	—	*	*
10a	spring 1965	8·1	530	31	9·9	590	43	43
3b	winter	9·2	160	46	9·0	190	82	80
2	winter	10·5	500	26	10·7	500	53	45
4c	summer	11·0	500	52	—	—	*	*
1c	winter	11·8	520	24	11·8	540	50	57
5	spring	12·1	520	26	12·5	540	41	42
7	autumn	12·6	190	26	12·2	174	52	53
10a	spring 1970	12·6	500	22	13·4	500	38	41
4b	summer	14·5	700	20	15·6	700	38	36
1b	winter	16·9	460	15	16·9	480	35	26

* no flowering
† the temperature at plant level was a little higher, see Table 4.3

In Table 4.2 the examples are selected to allow comparison of development at similar temperature but different radiation. The number of days from sowing to LS3 and from sowing to flowering for the same variety at several places is compared with the mean temperature and the radiation of each period. The places are arranged in order of increasing temperature.

High radiation and moderate heat loss. Two examples may be taken of pairs of localities with similar temperature but faster development at high radiation than at low. The first pair includes the plants at sites 10a (Alaska 1965) and 3b (Melbourne). The temperature is low (between 8° and 9°) but the radiation is high in the Alaskan summer and low in Melbourne after sowing in winter. In Alaska compared with Melbourne the barley takes 70 per cent of the time from sowing to LS3 while the pea and oat take only half as long from sowing to flowering.

The second pair includes the plants grown in winter at high altitude at site 2 (Mexico) and at site 3b. The temperature is low (close to 11°) but again the barley, pea and oat in the higher radiation in Mexico are nearly twice as fast.

High radiation and excessive heat loss. Two other pairs of examples indicate that high radiation is more or less counterbalanced by heat loss at very high altitude and latitude. The first pair compares plants grown at Timberline in California (site 4c) with the winter sowing at Melbourne (site 3b). In spite of higher radiation and a slightly higher mean temperature at Timberline, the length of the period from sowing to LS3 in barley is a little longer than at Melbourne. The second pair compares the plants grown at Pt Barrow, Alaska (site 10b), with the winter sowing at Melbourne. Here again, the higher radiation at Pt Barrow cannot compensate for the heat loss and although the temperature at plant level is raised somewhat (see Table 4.3b), the barley still takes longer than at Melbourne.

TEMPERATURE

The most desirable record of temperature in relation to plant development would be that of the tissues (shoot apex and leaves) directly concerned with the hormonal balances controlling the reproductive stages, but this was only available at two of the sites. However, the temperature in the standard meteorological screen with its base at about 1 m above ground has been shown to be relevant to days from sowing to flowering in the species under investigation (Aitken 1955a, 1966b). Even the soil temperature at 2 cm depth can be used to predict this period for maize in the warm humid growing season of Indiana in the United States (Wang 1958). The temperature was therefore measured daily by a maximum-minimum thermometer or by a thermograph in the standard screen, close to the plots at all sites. In addition, temperature was also recorded at plant level at three places (Chapingo, Mexico; College and Pt Barrow, Alaska) and within the shoot at two places (Melbourne and Pt Barrow).

Temperature is a powerful factor in plant development as shown in the following chapters. Various measures from the daily records may be used to show the relationship. The most important are the daily heat sum (mean daily temperature multiplied by mean daily photoperiod), the daily effective maximum temperature, the daily maximum and the daily mean (see Table 10.1). For simplicity the daily mean temperature is used as a rule throughout the book, usually in the form of the mean value for the relevant period of plant development.

Daily temperature

Temperature varies diurnally and seasonally throughout the year at the one place, the degree depending on latitude, altitude and type of climate.

During the 24 hours of the day the temperature curve is lowest at dawn and highest a little after midday. The amplitude and the mean will change from winter to summer, with the least differences in both measures at places close to the equator in a maritime environment.

The tropical island. The temperature regime of the Hawaiian Islands is illustrated by the sites 1a–c (Fig. 6.1). At site 1a, which is near sea level, the daily range of temperature is only 5° throughout the year, and the difference between the mean daily values for winter and summer months is even less

(3°). At the higher sites (1b, c), the daily range is wider but the seasonal difference between the mean temperatures is still very small because of the tropical situation.

Other environments. In contrast the temperature varies more within the day and between seasons in a continental climate and at higher latitudes and altitudes. Figure 4.5 gives samples of the daily temperature curves for three successive days, selected to show the trends in daily amplitude and mean through the growing season at the three different sites 2, 3b and 10a.

The figures for site 2 show the wide daily range of temperature in both winter and spring at Chapingo, Mexico. This occurs because of its position at high altitude inland and because of the low humidity in those seasons. The

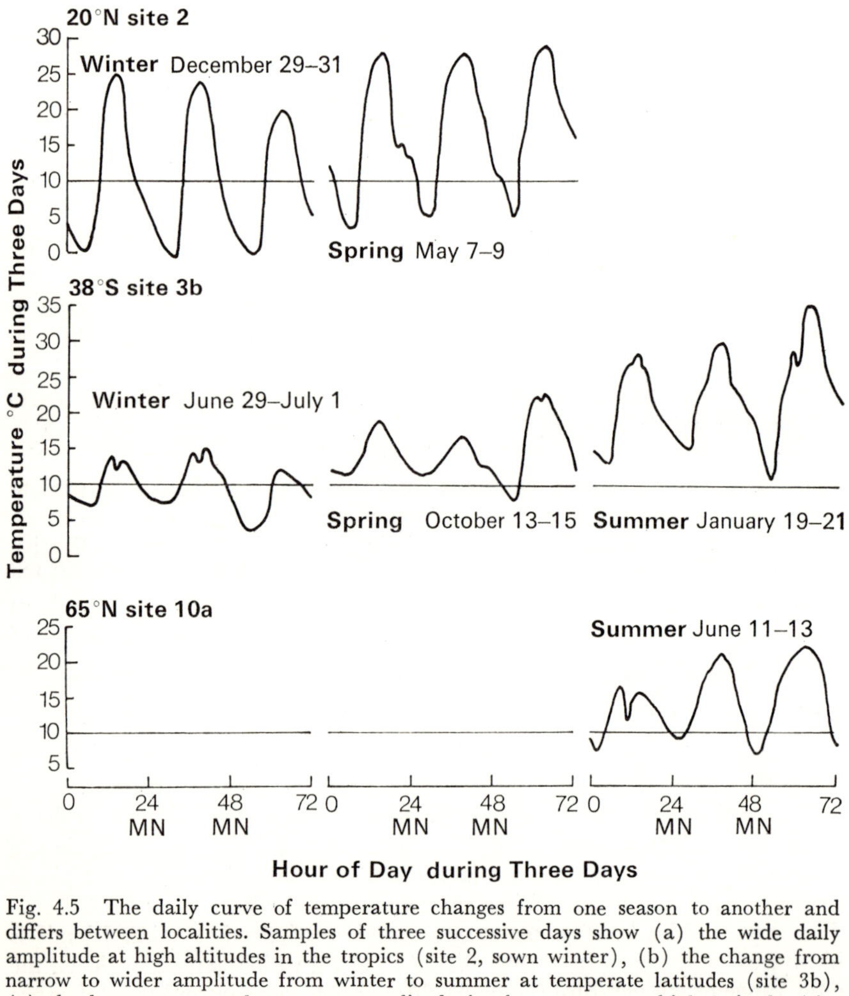

Fig. 4.5 The daily curve of temperature changes from one season to another and differs between localities. Samples of three successive days show (a) the wide daily amplitude at high altitudes in the tropics (site 2, sown winter), (b) the change from narrow to wider amplitude from winter to summer at temperate latitudes (site 3b), (c) the lower mean and narrower amplitude in the summer at high latitude (site 10a) compared with the same season in the previous sample (MN = midnight).

importance of the size of the land mass to the daily range of temperature can be appreciated by reference to the Hawaiian site 1c. This is comparable in altitude and latitude but is much more equable because of the small size of the island of Maui and the greater humidity (Fig. 6.1c).

In Figure 4.5 the examples from the growing season at Melbourne (site 3b) show first the narrow daily range of temperature and the mean close to 10° which is typical of many dull, moist winter days. The adjoining samples show that both the range and the mean will increase in spring and even more so in summer.

At site 10a the growing season is confined to summer. The daily range is less extreme than in summer at the other two places. The minimum is also reached more rapidly after midnight than at lower latitudes because of the continuous light.

Temperature at plant level

The temperature of the air close to the shoot tips of the plant may be expected to vary a little from that of the standard screen according to height of the plant above ground, the slope and surface of the ground, the spacing of the plants and the wind. Some species may have their shoot tips within 1 cm of the ground all their life as in the prostrate subterranean clover, or during their vegetative stage as in cereals, or only for a few days after emergence as in pea.

It is therefore of interest to note that the temperature close to the ground and in the low-growing plant is slightly higher than in the standard screen in some growing seasons with high radiation.

Temperature close to ground. Figure 4.6 gives examples of thermograph traces from sites 2 (Chapingo, Mexico) and 10a (College, Alaska) two weeks after the sowing. At College they show that the temperature within 6 cm of the ground is higher by about 2° throughout the 24 hours than in the standard screen. The higher temperature persists at this level for the first month after sowing, becomes less in the following month and disappears by August.

At Chapingo two sets of records are added to that in the standard screen. The screened box thermograph, at 1 cm on the bare ground of the plot, shows a higher temperature between the highest and lowest part of the diurnal curve which raises the mean value about 4° above that in the standard screen. The exposed metal rod at 5 cm over grass gives much more extreme curves than in the standard screen and a higher daily mean. The temperature remains higher during the first three months in both records.

Temperature within plant. The temperature within the plant compared with the standard screen varies according to site and time of year. Table 4.3 shows that the temperature in the shoot tip of subterranean clover and of ryegrass is slightly less at site 3b (Melbourne) in winter when the radiation is lowest but no different in autumn.

At site 10b (Pt Barrow) the radiation is high at the start of the growing season in July. Although the temperature at screen level is very low, at ground level (see also Bliss 1956, Sørensen 1941) and within the leaves of a

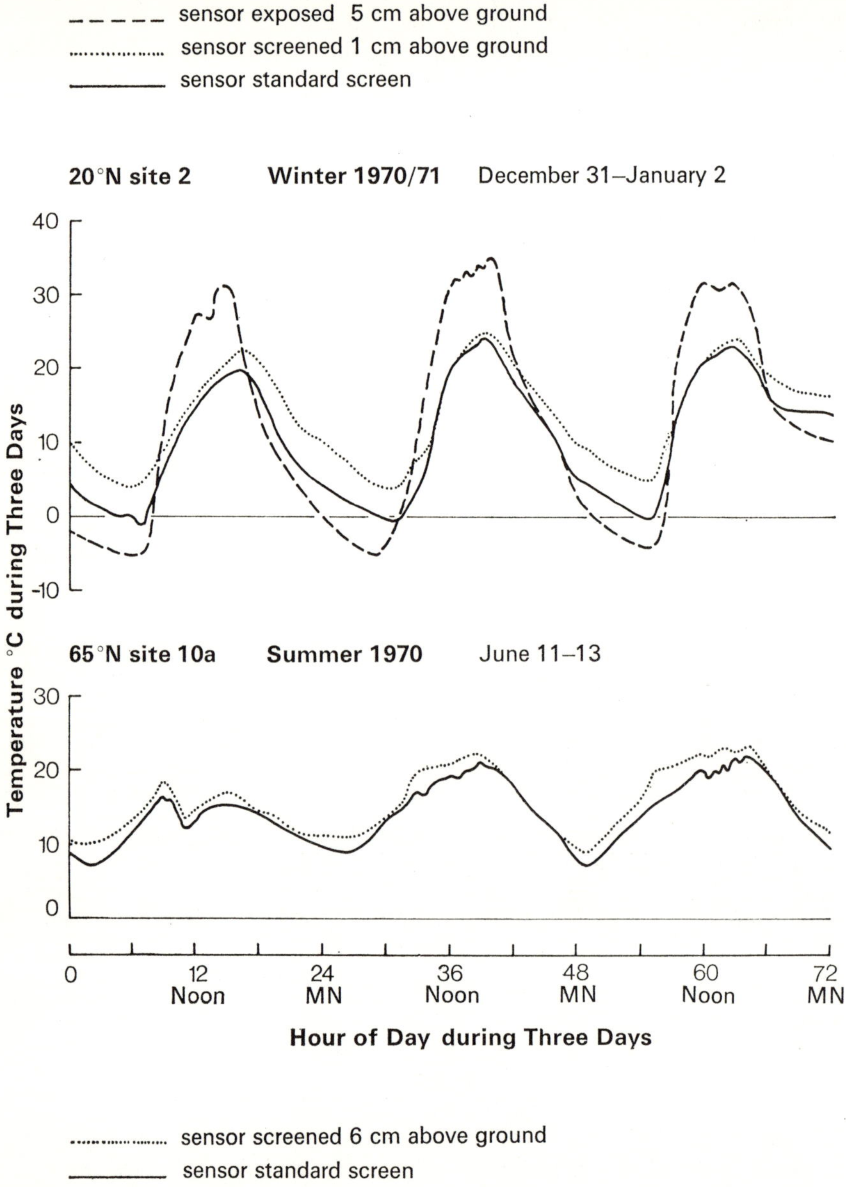

Fig. 4.6 At high altitudes and latitudes the temperature soon after sowing may be warmer close to the ground than in the standard screen. This is shown by the temperature curves for three successive days at site 2 in Mexico, sown winter and at site 10a in Alaska, sown late spring (MN = midnight).

Table 4.3

The differences between the temperature within the plant and in the standard screen at moderate to low radiation and at high radiation

Moderate to low radiation, site 3b, Melbourne

Season, period and measures	Temperature in shoot tip compared with screen			
	subterranean clover		ryegrass	
	mean difference °C	no. of samples	mean difference °C	no. of samples
Autumn, 12–22 Apr				
radiation 220 cal/cm²/dy	0·1	15(day)	0·1	8(day)
range of temperature during period 8–25°C	−0·3	7(night)	0·1	3(night)
Winter, 6–29 June				
radiation 138 cal/cm²/dy	−2·0	9(day)	−1·8	7(day)
range of temperature during period 4–14°C	−1·5	7(night)	−0·6	3(night)

*High radiation, site 10b, Pt Barrow**

Season, period, radiation	Mean daily temperature standard screen, at 10-day intervals		Temperature gain over that in standard screen		
	first date	°C	ground °C	grass leaf °C	*Ranunculus* leaf °C
Summer, 25 June to 3 Sept	25 June	3·0	2·5	2·4	4·3
	5 July	5·0	2·7	3·1	3·9
radiation	15 July	1·9	3·1	3·1	4·6
July 520 cal/cm²/dy	25 July	2·3	3·0	3·5	4·7
Aug 230 cal/cm²/dy	4 Aug	2·0	3·2	2·4	2·9
	14 Aug	1·6	3·4	3·2	3·2
	24 Aug	1·2	3·6	3·9	3·9

* Brown 1970 and pers. comm.

grass and a dicotyledon (*Dupontia fischeri* R. Br. and *Ranunculus nivalis* L.) it is higher by up to about 4°. Thus growth becomes possible for the few weeks of summer.

PHOTOPERIOD

The photoperiod (hours of light per 24 hours) is related to rate of development in two ways: through its effect on the daily heat sum and through its effect on the photoperiodic responses of sensitive varieties.

Daily heat sum

The photoperiod is connected with the daily sum of temperature-hours as it affects the rates of leaf appearance and seed ripening. It was shown in the previous chapter (p. 36) that the rate of leaf appearance increases with the mean daily temperature when the plant is grown at constant temperature and the same radiation. In the field, however, the diurnal temperature varies in amplitude, in mean and in the number of hours above the mean. These values are related to the photoperiod through the daily radiation.

In consequence the product of the mean daily temperature and photoperiod in the field is even more closely related than the mean temperature to the rate of development of the simplest early-maturing varieties. This is shown by the high negative correlation between this product and the length of the growing period in the very early varieties of pea and oat (Table 10.1).

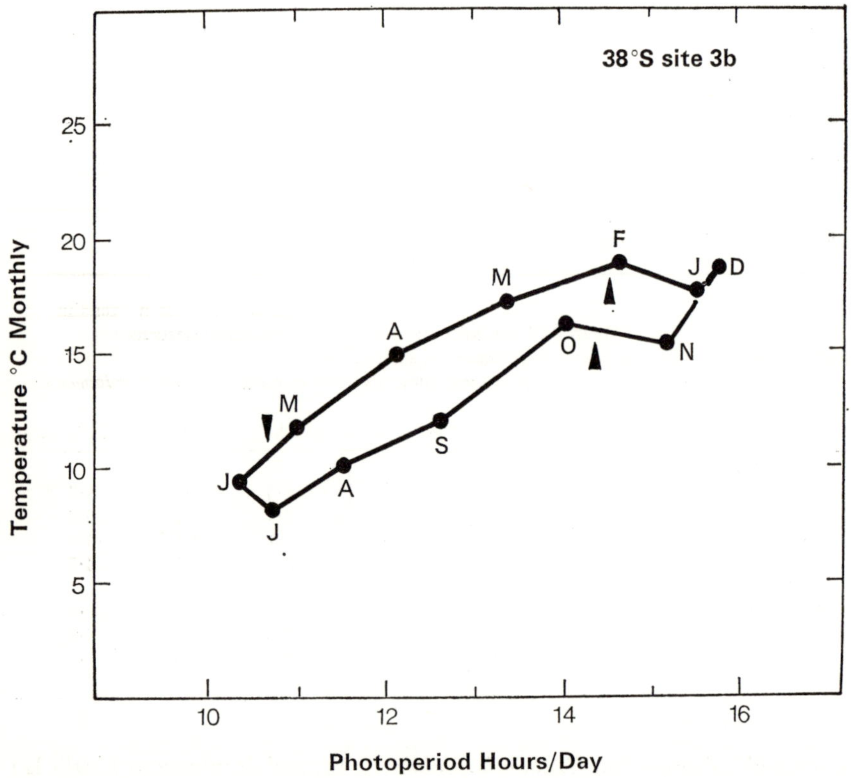

Fig. 4.7 Depending on the time of sowing (see arrows) and location, the plant may grow in different temperatures and photoperiods as shown by the mean daily values of each factor in successive months of the year beginning in 1963 at site 3b in Melbourne, Australia. Sown in late summer (17 Feb) the plant grows at high but declining values; sown at the end of autumn (30 May) the plant begins to grow at low values which then rise; sown in mid-spring (25 Oct), growth proceeds at higher values. Each dot represents the mid-monthly values of daily temperature and photoperiod (including civil twilight).

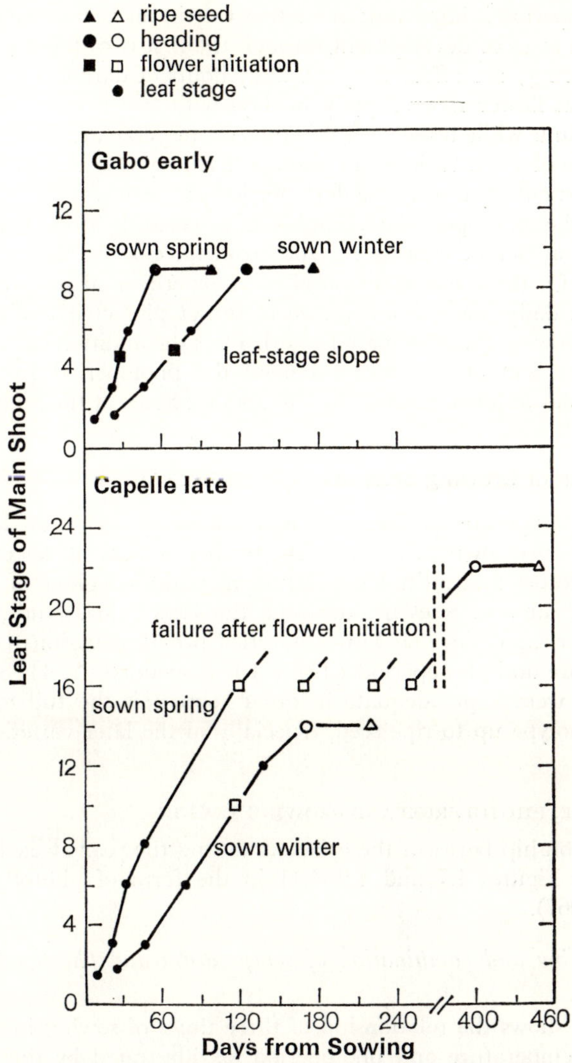

Fig. 4.8 Time of sowing affects the progress of wheat in leaf stage and in development with days from sowing, as shown by an early and a late wheat sown in spring and in winter at site 3b in Melbourne. The leaf stage is shown at intervals up to heading, and the continuous line indicates the leaf-stage slope up to heading and the time taken from heading to ripe seed. In the early variety, both the rate of leaf appearance and of development are faster after sowing in spring than in winter. In the late variety the rate of leaf appearance follows the same pattern. The rate of development, however, is much slower after sowing in spring than in winter because of death of the first-formed reproductive shoots.

Varietal sensitivity to photoperiod

The photoperiod is important in another way in the more complex varieties in which the rates of development through the first two stages (sowing to FI and to flowering) are affected by the daily hours of darkness (nyctoperiod). Varieties that flower more rapidly in days with few hours of darkness are 'long day' plants while those with the converse response are 'short day' plants. The response of some varieties to photoperiod may also be affected by temperature, particularly during the dark portion of the day.

Because the photoperiodic stimulus is affected by low light intensity, down to about 5 foot candles for subterranean clover (Aitken 1955a), the photoperiod for the growing season of each experimental site was calculated as the mean daily hours from sunrise to sunset plus civil twilight for each month after sowing (Smithsonian Tables). Thus the addition of civil twilight* before sunrise and after sunset extended the photoperiod from about 45 minutes per day at low latitudes (20°) to about 2 hours at the higher latitudes.

3 Selection of Growing Seasons

The nine temperate species, each represented by early and late varieties (Table 2.1), were sown in a range of growing seasons by selecting sites in different latitudes and altitudes to represent various regions of agricultural and pastoral interest. Sites in Australia, the Hawaiian Islands, North and Central America, Wales and Sub-Antarctica provided various combinations of temperature and photoperiod (Table 4.1, Appendixes 2, 4). Soil moisture and fertility were kept adequate in order to permit the full expression of maturity genotype up to ripe seed, especially in the later varieties, wherever possible.

TEMPERATURE, PHOTOPERIOD AND GROWING SEASON

The relationship between these factors during the year at each location is presented in Figures 4.7 and 4.9–4.11 in the form of photothermographs (Ferguson 1967).

Time of sowing and combination of temperature and photoperiod at Melbourne

Figure 4.7 shows the relationship of three times of sowing in a temperate latitude to temperature and photoperiod, as illustrated by the mean daily values at Melbourne, site 3b in 1963–4. Monthly temperatures and photoperiod are associated but the temperature is lower in the spring compared with the autumn although the photoperiod is similar. The three sowings therefore were followed by different types of growing season. The plants sown in winter grew at the lowest level of temperature and photoperiod of the year for some months, compared with those sown in summer and in spring. These two growing seasons had a similar photoperiod at first but plants sown in summer began growth at temperatures about 2° higher than plants sown in

* Civil twilight is the period from the time the sun is 6° below the horizon to sunrise, and from sunset to 6° below the horizon.

spring. The season became cooler and the photoperiod lower as the season progressed. The conditions were reversed for the spring-sown plants.

The development of a variety is easily related to the relevant climatic values. Thus Gabo wheat when sown in spring (25 October) took 100 days from sowing to ripe seed (Fig. 4.8). During this time (Fig. 4.7) the mean monthly temperature of the growing period rose rapidly from 15° to 17° and remained at this level while the photoperiod rose from 14.5 to 15.7 hours by late December and then fell to 15 hours by the end of January.

The same variety took 180 days when sown in early winter (30 May). During this time the mean monthly temperature was about 9° for the next three months and then rose to 15° by mid-spring (October). The photoperiod rose slowly from just over 10 hours in the first month to 14 by mid-spring.

Seasonal changes in temperature and photoperiod in a range of experimental sites

In Figure 4.9 the graphs are arranged with the highest latitude to the right. They show two trends: (a) to lower mean temperatures with increased

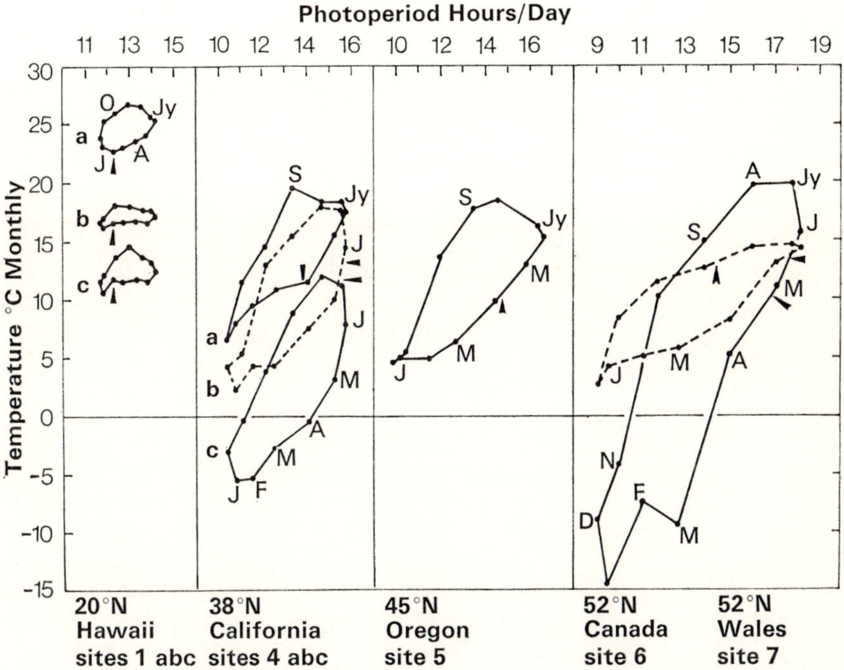

Fig. 4.9 Both latitude, altitude and time of sowing (see arrows) affect the temperature and photoperiod (including civil twilight) in which the plant grows. This is shown by the mean daily values of each factor in successive months in 1963 (see Table 2.2 for altitude and Appendix 2 for dates of sowing). At low altitude the values for temperature and photoperiod for the same season decrease with increasing latitude but much more strongly in winter than in summer. With higher altitude the same time of sowing is followed by lower temperature, as shown in winter at sites 1a-c in Hawaii and in spring at sites 4a-c in California.

latitude and altitude and (b) to an increased range within the year of both temperature and photoperiod with latitude. In addition an example of the range of monthly temperature during the year in maritime and continental climates can be seen by comparing sites 7 (dashed line) and 6 respectively.

The type of growing season at each location is assessed by noting first the position of the sowing date in the particular graph and then the temperature and photoperiod of the following months. Taking the hot growing season at site 1a in Hawaii as an example, it can be seen that the plants sown on 21 February in the late winter grew at high temperature and moderate photoperiod over the next months. Both factors increased only to a small extent from late winter to summer. Plants sown at site 1c in winter grew at lower temperature each month but the same photoperiod.

In Figure 4.10 sites 2 and 10a are fairly similar in temperature during the early part of the growing season but differ greatly in photoperiod. This con-

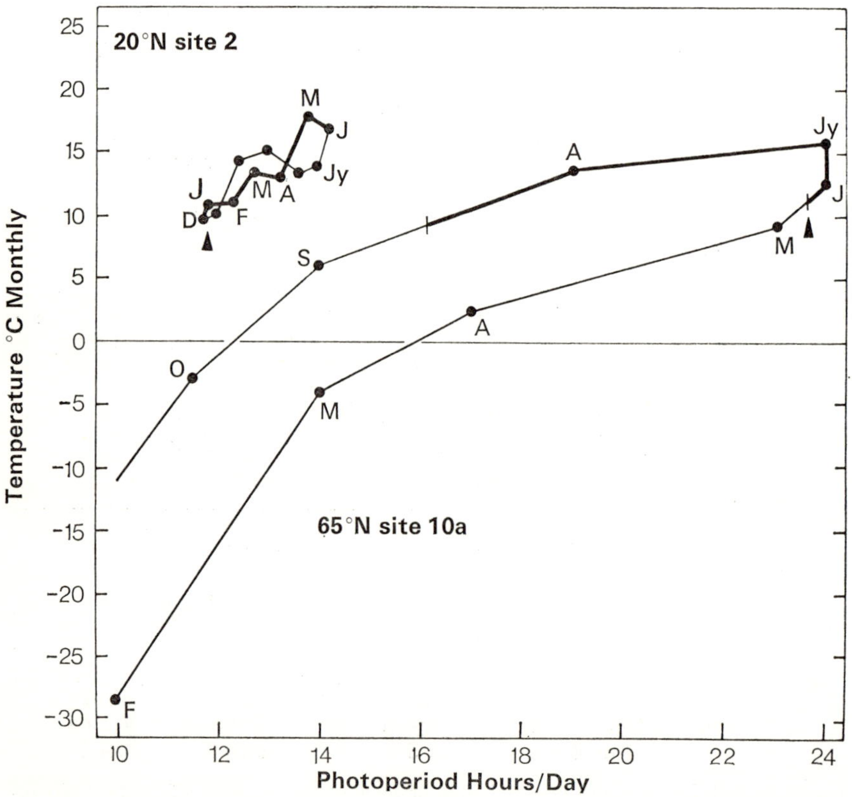

Fig. 4.10 The location of sites in different regions can provide growing seasons that differ strongly in one climatic factor. This is shown by the moderate photoperiod that follows sowing in winter at site 2 in Chapingo, Mexico, compared with the continuous light that follows sowing in spring at site 10a in College, Alaska. Each dot represents the mid-monthly values of daily temperature and photoperiod (including civil twilight) for the particular month. Each sowing date in 1970 is shown by an arrow, and the monthly values of the growing season (thick line) are read from the arrow.

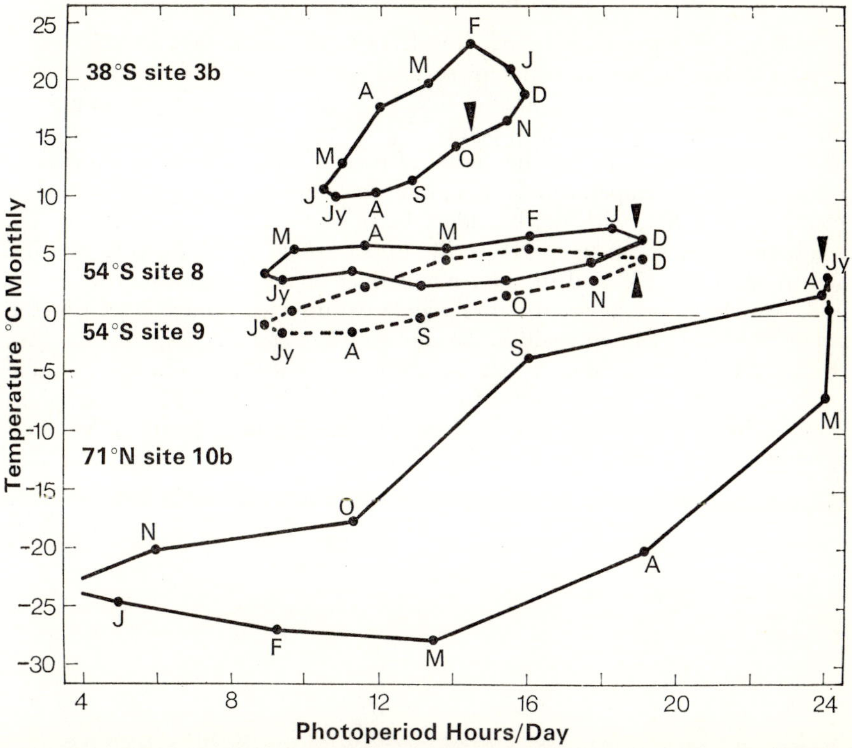

Fig. 4.11 The location of sites at higher latitudes provides spring-summer growing seasons with a trend to longer photoperiod but lower temperature, as shown by comparing these climatic factors in successive months of the year at site 3b (Melbourne 1968), site 8 (Macquarie I.), site 9 (South Georgia I.) and site 10b (Pt Barrow). Each dot represents the mean values of daily temperature and photoperiod (including civil twilight) for the particular month. The arrow indicates the sowing date at each place, and the monthly values of the growing season are read from the arrows in an anti-clockwise direction.

trast was possible by sowing the plots in winter at high altitude in Mexico and by sowing in late spring in central Alaska. In both cases, irrigation was needed and was applied throughout the growing period of the varieties in Mexico and during early growth of the plots in Alaska.

Figure 4.11 compares the conditions at Melbourne (site 3b) with three very cool growing seasons at high latitude. Sites 8 and 9 were on Macquarie and South Georgia Islands, and site 10b on the northern coast of Alaska at Pt Barrow. On the two islands the mean monthly temperature varied little (Table 4.1) during the year but the photoperiod varied a lot. At Pt Barrow the range of mean monthly temperature is large but the peak of the monthly curve only reaches about 3° in mid-summer even though the photoperiod is continuous. However, note the higher temperature at plant level (Table 4.3).

Flowering Time, Climate and Genotype

A CLASSIFICATION OF GROWING SEASONS

Although the previous figures (4.9–4.11) have shown that some growing seasons differ greatly in one factor such as temperature at three altitudes in Hawaii, better comparisons are possible if the records are grouped in a series on the basis of temperature. Appendix 4 gives three measures of temperature and one of photoperiod for the period of three months after sowing. Whichever measure of temperature is used, the growing seasons may be grouped as cool, warm or hot, with the boundary between the first two ill-defined. On the basis of the mean temperature the upper boundary of the cool growing seasons could be 13° and the upper boundary of the warm group would be 17°. The mean photoperiod varies widely within each group as seen in the range from 11 to 24 hours within the cool growing seasons.

The growing seasons can also be grouped on the temperature and the photoperiod of the first month after sowing (Appendix 2). These groupings may be important to the rate of development of certain kinds of maturity genotype, such as those delayed by temperature above a certain threshold (p. 114).

V Concept and Properties of Maturity Genotype

What controls the time of flowering and of seed ripening in the one variety in different climates? This is not only a practical problem for both farmer and adviser but also part of the wide field of developmental and ecological genetics which so urgently needs investigation (Clausen 1958b).

1 Concepts of Maturity Type (MG)

Over the last two hundred years views on the relationship of the plant's growing period and climate have changed greatly. Initially it was thought that temperature, between the lethal limits of cold and heat, was the chief factor directly and positively controlling the rate of development of a species (Abbe 1905, Maclagan 1933). Thus the growing period would be shorter in warm than in cool climates.

The real picture is much more complex. In 1918 Gassner demonstrated that *low* temperature could hasten the development of certain late-maturing types of plants such as winter cereals and biennial crucifers, and this has been confirmed by other workers. Day length was also found to affect various species (Garner 1920): long days hastened the development of varieties within some species but retarded those of other species.

Botanically it is now more widely realized that genetic variation in time of flowering is great and occurs both between and within species. In an annual species such as pea, even the latest-flowering phenotype is relatively fast compared with the earliest form of the woody perennial fruit trees (e.g. apple, plum) which are more slowly developing species (Kozlowski 1971). Vavilov (1922) found that each of many herbaceous cultivated species included a wide range in maturity character.

The first step in defining the genetic basis of maturity character was the realization that the vegetative stage ended at FI, not at first flower. The time of FI was therefore a key event which could determine the time of flowering and seed ripening (subterranean clover, Aitken 1941; wheat, Aitken 1966a; ryegrass, Cooper 1952).

Another step was taken when two concepts were proposed for the genetic control of maturity character. One of these was control by differences in the *tendency to flower* which resulted in minimum differences in vegetative nodes between varieties sown at the same time. Purvis (1934) drew attention to the lower number of minimum vegetative nodes in early varieties of a species than in late ones, as shown in rye. The second concept was that of control by differences in *flowering responses*. The varieties with responses less inhibited by high temperature and/or less sensitive to day length reached FI earlier and thus flowered earlier than those with more reactive responses, as found in ryegrass, *Lolium* spp. Cooper 1952, 1956ab) and subterranean clover (Aitken 1955a, Evans 1959).

Both concepts have been helpful, but neither can be used to predict the growing period of a variety in an environment different from that in which it has been actually grown and classified as early- or later-flowering. Maturity character is determined not only by the genetic tendency to flower but also by other factors which control the rates of development through all stages to ripe seed.

2 Present Concept of MG

In spite of the complexity of the interaction between the MG of a variety and the climatic factors in the field, the development analysis of the same varieties in different growing seasons on a world basis, as reported in this book, has provided a more comprehensive, third concept of MG. Such an approach shows that the growing period of a variety results from the interaction of two properties of its MG with the environment. The first property is the tendency to flower as indicated by the minimum number of vegetative phytomers at FI (minimum VN). The second property is a group of responses to the environment of each growing season (Fig. 5.1).

At this point the present concept could be objected to as lacking in chemistry or in statistics. Presumably both the tendency to flower and each of the individual responses represents a hormone still to be identified (Evans 1969). This implication must be left to the biochemist. The urgent task is to reduce to order the mass of information on reproductive development which may appear chaotic as illustrated in Figure 10.1. A more statistical approach can only show broad relationships of rate of development with temperature and photoperiod as indicated in Table 10.1. The analysis of genetic control into a set of reactions makes possible an understanding of the expression of different kinds of MG according to climate.

The interaction of MG with environment is expressed through the three components of development: VN, rate of leaf appearance and rate of seed maturation. Once the MG can be identified from the minimum VN and from the various plant responses and their intensities, its total expression as growing period can be predicted for a particular type of growing season.

3 First Property of MG: Tendency to Flower

The tendency to flower can vary greatly both within and between species. It affects the timing of FI and is indicated by the minimum VN produced in the most favourable environment for the change to the reproductive stage.

Figure 5.1 gives examples of the formulae that represent early and late varieties of two species. In oat the late variety has a higher minimum VN and more responses than the early one. In subterranean clover the relationship of the late variety to the early one is the same as in oat as far as the first property is concerned but different with regard to responses.

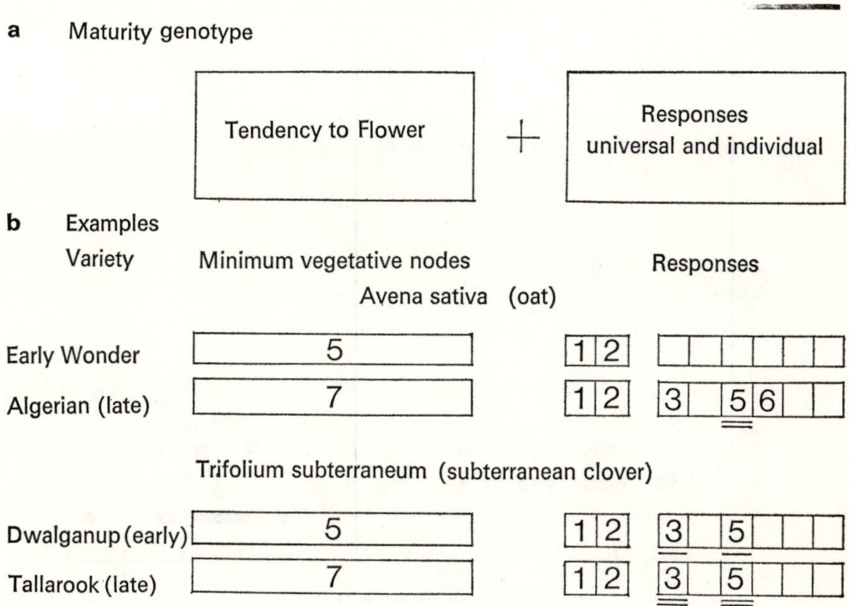

Fig. 5.1 (a) The concept of the MG, which includes two properties, the tendency to flower and a group of responses (see Table 5.2) that also affect the rate of development. (b) Two ways in which early and late varieties of a species differ. The late variety reaches flowering more slowly in the oat species because of the individual responses, and more slowly in subterranean clover because two of these responses are more intense, as shown by the double bar compared with the single bar below the response index.

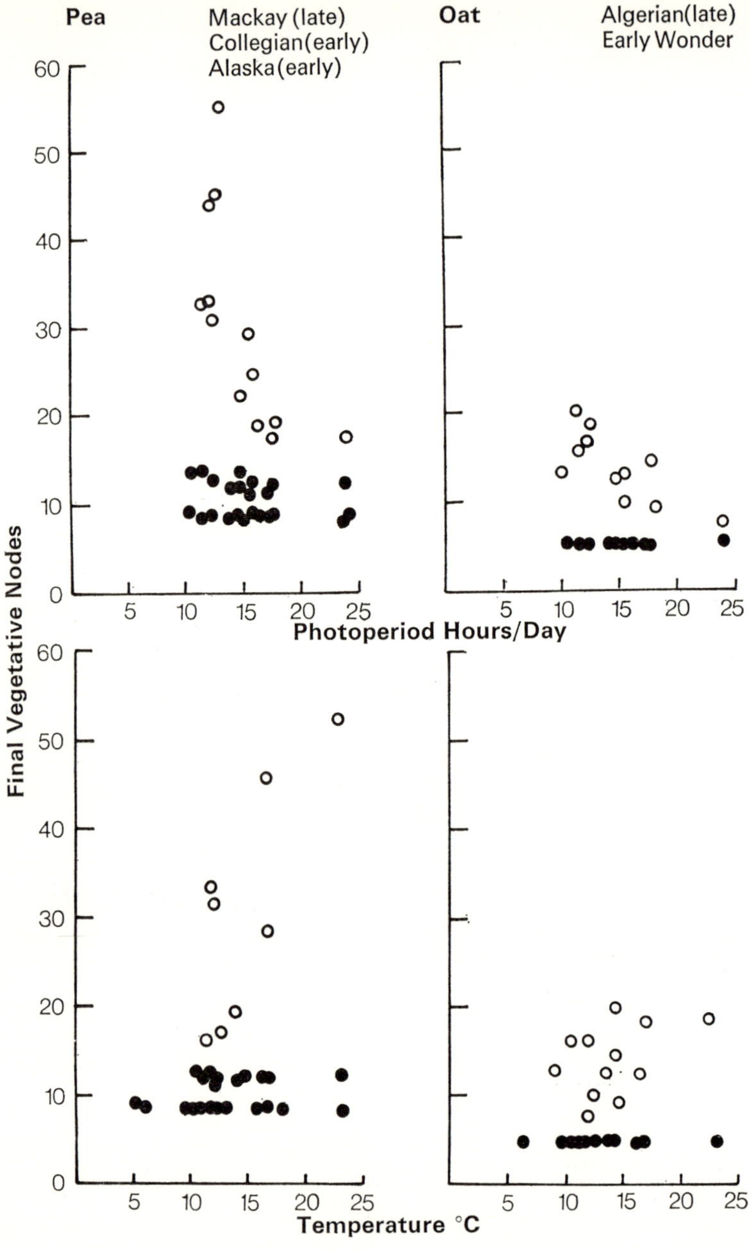

Fig. 5.2 When grown at the various climate sites, the MG of a variety affects the relationship of VN with the mean temperature and photoperiod of the vegetative stage (sowing to flower initiation). In some species the VN of a variety is unassociated with either factor as shown by the stable value in the early peas and the early oat. In other species the VN is closely associated with one factor—photoperiod in the two red clover

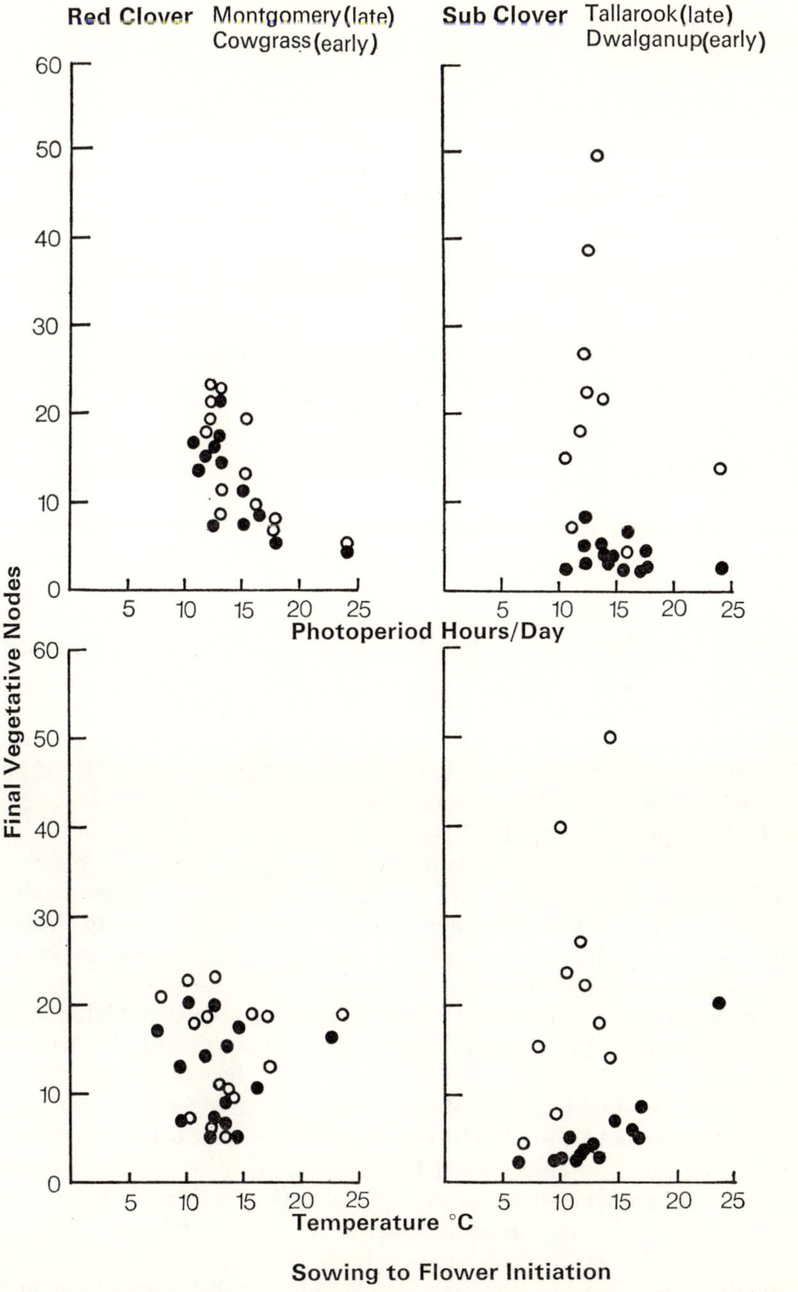

varieties, and temperature in the early subterranean clover. The four examples show some of the differences in pattern of VN between early (solid circle) and late (open circle) varieties within and between species. Each circle represents the mean of ten plants, the values normally with a coefficient of variation 2 per cent or less. The VN is measured on the main shoot in pea and oat and on basal laterals in the other species.

In some early varieties of species such as pea and oat, progress towards FI is so far advanced that the number of vegetative phytomers remains the same, whatever the environment. Thus the minimum VN would seem to be determined before germination. In other varieties, particularly those with a late MG, assessment is more difficult because FI is usually more or less delayed by genetic responses to the growing season. The most favourable environment for expressing only the minimum VN must therefore be found and defined precisely for such an MG.

VARIATION WITHIN AND BETWEEN SPECIES

A large range in MG within a species is likely to be based on both a decreasing tendency towards FI and on developmental responses that may further reduce the tendency to flower in some environments. Figure 5.2 gives examples from early and late varieties of four species. In each variety the VN at a given site is placed along the vertical scale and the mean temperature (lower section) or the photoperiod (upper section) of the vegetative stage at the same site is placed along the base line. Each symbol indicates the mean value of the variety in one growing season.

For instance, the highest VN (54) in the late pea variety Mackay occurs at site 1a in Hawaii. The symbol is located in the upper section above the value of 13-hour photoperiod and in the lower section above the value of 23°. The position of the symbols in relation to each of the factors indicates the degree to which the minimum VN is obscured by developmental responses. The fact that the VN remains stable in a range of values of a climatic factor indicates the minimum VN characteristic of the simplest MG in early varieties. A trend of values indicates that the MG includes a developmental response to the factor concerned.

In both pea and oat the VN in the earliest varieties is small and unchanged in a range of both temperature and photoperiod. The two early varieties of pea, Alaska and Collegian, reveal a stable or almost stable tendency to flower. The earlier one has a minimum VN close to 9 and the slightly later one has a VN of about 12 at all the sites. The early variety of oat has a minimum VN of 5.

In the same two species the VN for the late varieties is highly variable. The late pea Mackay reaches its lowest VN (17) at site 10a in Alaska with a combination of 24-hour photoperiod and 12° temperature, and the VN tends to increase with shorter photoperiod and with higher temperature. The late oat Algerian also reaches its lowest VN in a similar combination, but the scatter of values is only related broadly with photoperiod. The minimum VN in the late oat is closer to that of the earliest variety (7 compared with 5) than is the case with the late pea variety (17 compared with 9).

In each of the two other species, both early and late varieties contain genetic responses to temperature and/or photoperiod in addition to their minimum VN, as shown by the scatter of VN values in relation to each climatic factor. In red clover the values of minimum VN for the two varieties coincide at VN5 on the lateral in a 24-hour photoperiod and at about 12°. Both varieties respond to photoperiod, with VN increasing with shorter photoperiod.

In subterranean clover the pattern of variation is different. The early variety has a minimum VN of 2 on the lateral over a range of photoperiod and with comparatively low temperature (between 6° and 13°). In contrast to the early varieties of pea and oat the VN tends to increase with temperature. The late variety has a minimum VN of 5 which is confined to one place with long photoperiod and low temperature. This combination results from the early summer sowing at site 8, Macquarie Island. The minimum VN in the late subterranean clover compared with that of the early one is therefore 5 and 2 when measured on the lateral and the difference between the varieties resembles the relationship in the oat rather than in the pea species. In contrast to the other late varieties, the VN values appear unrelated to both climatic factors.

TREND OF MINIMUM VN WITHIN A SPECIES

In Table 5.1 six pea varieties of different MG provide evidence of several differences in the tendency to flower within a species. The later varieties are also more sensitive to photoperiod. The table shows the effect of locality on the expression of MG in four places differing mainly in photoperiod. The three measures of maturity are (a) the VN, (b) the days from sowing to FI (stage 1), and (c) the days from sowing to flowering (stages 1 plus 2).

Because of the combination of continuous light and moderately low temperature, the Alaskan growing season allows the most rapid development to FI of the later-maturing peas. Hence the range in VN indicates three levels in tendency to flower. The most rapid tendency is shown by a mean VN of 8·6 and is found in the Alaska variety. The slightly slower tendency results in a mean VN of about 12·5 and is found in the next two varieties, Collegian and Canner's Perfection. The slowest tendency is expressed by a mean of about 17 and is found in both the mid-season variety and the two late ones, Mackay and Austrian Winter.

The differences in the tendency to flower affect the length of the vegetative stage which increases from 11 days in the early Alaska variety to 29 in the three latest varieties, a three-fold increase. In consequence the number of days from sowing to first flower rises from 38 in the earliest variety to 55 in the latest. The full effect of the differences in minimum VN is offset by growth of the later varieties in higher temperature than is available for the earlier varieties.

4 Second Property of MG: The Group of Developmental Responses

Analysis of the interaction of variety with growing season shows that up to eight responses (R) can affect the growing period of the species investigated. These are summarized in Table 5.2. Six are associated with temperature (R1, 2, 3, 4, 7, 8) and two with increased day length (R5, 6). A response that increases rate of development within a stage, reduces its length and is termed positive. A negative response reduces rate of development, tending

Table 5.1
The relationship of maturity group in six varieties of pea with the temperature and photoperiod of four growing seasons, as shown by three development measures

	3b Melbourne sown winter	3b Melbourne sown spring	6 Canada sown spring	10a Alaska sown spring
Growing season*				
Temperature °C:				
stages 1 and 2	10·6	17·0	15·7	14·1
Photoperiod hr/dy: stage 1	11·4	15·5	17·9	24
Variety and maturity	Measures of development†			
	Days for development stage 1			
Alaska early	13	6	10	11
Collegian early	31	17	24	17
Canner's Perfection early	41	20	29	21
Dun mid	76	38	35	29
Mackay late	118	48	35	29
Austrian Winter late	145	50	34	29
	Days for development stages 1 plus 2			
Alaska	80	33	35	38
Collegian	98	41	41	45
Canner's Perfection	105	46	47	51
Dun	135	59	57	57
Mackay	161	72	55	55
Austrian Winter	170	70	54	55
	Number of vegetative nodes per shoot (VN)			
Alaska	9·0	8·6	8·5	8·6
Collegian	13·7	13·5	12·3	12·2
Canner's Perfection	14·1	13·3	12·6	13·0
Dun	22·0	23·2	19·5	18·6
Mackay	32·4	26·4	19·3	17·0
Austrian Winter	33·2	25·0	18·4	17·6

* mean values for climatic factors are related to the time taken by the latest varieties
† mean values are from 10 plants per variety; coefficient of variation usually less than 1 per cent

thus to increase the length of a stage. Some of the responses vary in intensity according to species and MG.

The effect of several responses acting at the same time on the rate of development within a stage depends on their positive or negative nature and on the relative intensity of the responses. The addition of a weak negative response to a strong positive one could merely lengthen the stage a little whereas the reversal of the strengths could lengthen it enormously.

UNIVERSAL RESPONSES AND GENERAL DEVELOPMENT RATE

The positive relationship of temperature to development rate is based on the two responses R1 and R2 which are common to all varieties within and between species. In temperate species the positive relationship operates

Table 5.2
A classification of the plant responses that affect the rate of plant development, in relation to expression, type, intensity and occurrence in species

Index	Expression of response	Type	Intensity*	Occurrence
	Universal responses			
R1	increases rate of leaf appearance with increasing temperature	temperature-positive	+++	all temperate species tested
R2	increases rate of seed development with increasing temperature	temperature-positive	+++	all temperate species tested
	Individual responses			
R3	retards flower initiation: more vegetative nodes form	temperature-negative	+ ++ + ++ ++ +++ ++ +++ +	pea cereals ryegrasses subterranean clover red clover
R4	retards rate of development after flower initiation: slows leaf appearance and development of inflorescence	temperature-negative	+ ++	cereals
R5	hastens flower initiation: fewer vegetative nodes form	photoperiod-positive	+ ++ + ++ +++ ++ +++ ++ +++ + ++	cereals pea red clover ryegrasses subterranean clover
R6	hastens development after flower initiation: increases rates of leaf appearance and inflorescence development	photoperiod-positive	+ ++ ++	cereals ryegrasses
R7	rate of leaf appearance differs between species at same temperature	species modifies R1	+ ++ +++	slow to fast species
R8	increases rate of seed development at low temperature	variety modifies R2	+ ++	slow to fast varieties of wheat, pea

* in a species with a range of intensity in an individual response, the later variety tends to have the more intense form
+, ++, +++ responses absent to weak, strong, very strong

over a range of temperature which begins at a lower threshold than for tropical species. The first response, together with the VN, controls the lengths of the first two stages of development if no contrary responses are present. Similarly the second response, together with the number of reproductive nodes, controls the third stage.

Universal response R1

The first response affects the first and second stages of development by increasing the rate of leaf appearance as temperature rises between a lower and an upper boundary suitable for plant growth. The response affects the number of days from sowing to LS1 and to further stages (Plate 3A). It may

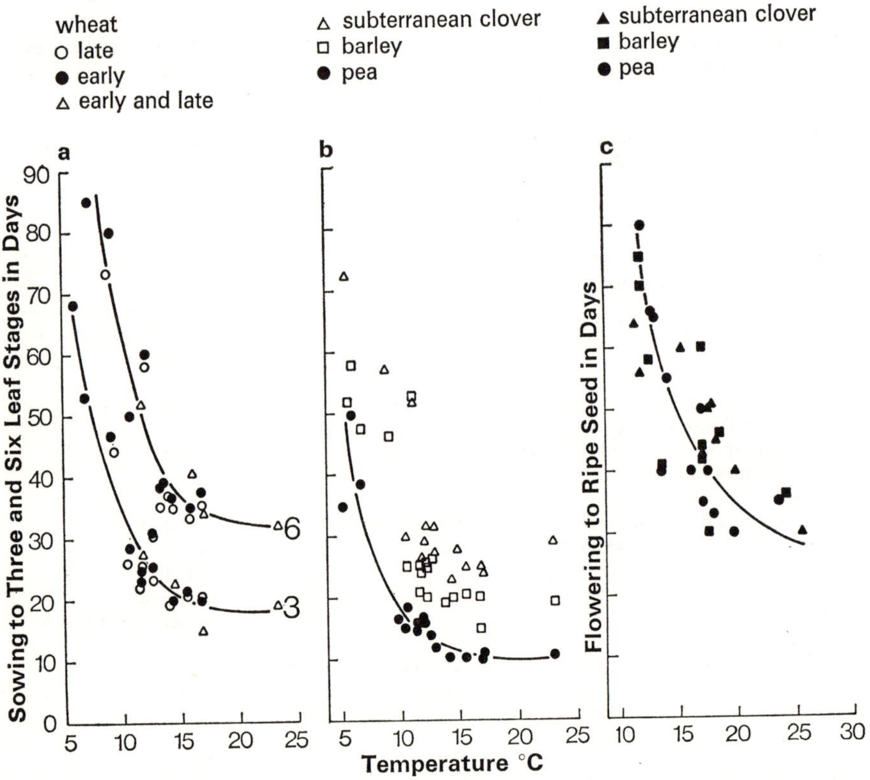

Fig. 5.3 The climate sites provide the following examples of the two universal responses to temperature. (a) Within a species, increased temperature, through R1, shortens the period from sowing to LS3 and LS6 irrespective of MG as shown in wheat with mean values for early, late and the two groups combined. (b) Between species the same trend occurs with the days from sowing to LS3 but in parallel curves because of intrinsic differences in rate of leaf appearance at the same temperature. This is shown by the lower position of the curve for pea compared with the trends for barley and subterranean clover. (c) Increased temperature, through R2, shortens the period from flowering or heading to ripe seed, irrespective of species as shown by the general trend for the three species. Each symbol represents the mean of ten plants at one site and the values have usually a coefficient of variation of less than 2 per cent.

also be measured by the number of days taken for each new leaf stage (days/leaf).

The relationship holds for the same species in both constant (glasshouse) and fluctuating (field) mean daily temperatures. This may be illustrated by comparing Figures 3.11 and 5.3ab. In the first case the evidence comes from varieties in the vegetative stage in order to avoid possible interactions with other responses. A comparison of wheat varieties shows a similar curve for the line depicting the change in length of the period from sowing to LS3 with temperature. Both curves indicate a faster rate of leaf appearance as the temperature rises between 7° and 17° and a much slower rate above this level. Pea also has the same shape of curve as wheat although its rate of leaf appearance is faster.

Universal response R2

The second response affects the third stage of development and is illustrated from plot evidence. The length from first flower to ripe seed is reduced to about one-third with increased temperature from about 12° to 25° as shown in Figure 5.3c for pea, barley and subterranean clover and in Figure 10.4c for wheat. In contrast to R1 the rate of seed ripening increases steadily between 17° and 25°. The trend is remarkably similar considering the difference between species in size of seed (50-fold between subterranean clover and pea) and in type of inflorescence (terminal or axillary).

INDIVIDUAL RESPONSES AND DIFFERENCES IN DEVELOPMENT RATE

In combination with the two universal responses, six others cause differences between varieties in length of growing period, through their presence, identity and intensity. These responses (Table 5.2) affect varieties both within and between the species under discussion. However, further types of response can be expected when other species are fully investigated. For instance, it is well known that species such as soybean, sorghum (*Sorghum vulgare* Pers.), maize, tobacco and chrysanthemum are in the 'short day' group—that is, they each include a negative response to photoperiod within their range of varieties (Garner 1920, 1923; Laurie 1930).

Of the six responses to be considered now, two are temperature-negative (R3, R4), two are photoperiod-positive (R5, R6) and two (R7, R8) modify the universal responses (R1, R2). All but R8 influence the lengths of the first two stages.

Negative responses to temperature

These operate before or after FI according to the individual response and reduce the rate of development with higher temperature.

Individual response R3. This response affects the first stage of development because it retards FI, thus increasing VN. It is present in all the species under discussion but more noticeably in subterranean clover and the two ryegrasses, and more in the later varieties than in the early ones.

Since universal R1 is in all varieties of a species, the presence of R3 in one variety is most easily demonstrated by comparing its development with that of another which is free of the extra response. Figure 5.4 shows the relationship of species and MG at three temperatures in the Hawaiian Islands (sites 1abc) to the length of the period from sowing to flowering (lower section of figure) and to VN (upper section). The early varieties of pea are free or almost free from R3 and therefore reach FI more rapidly with in-

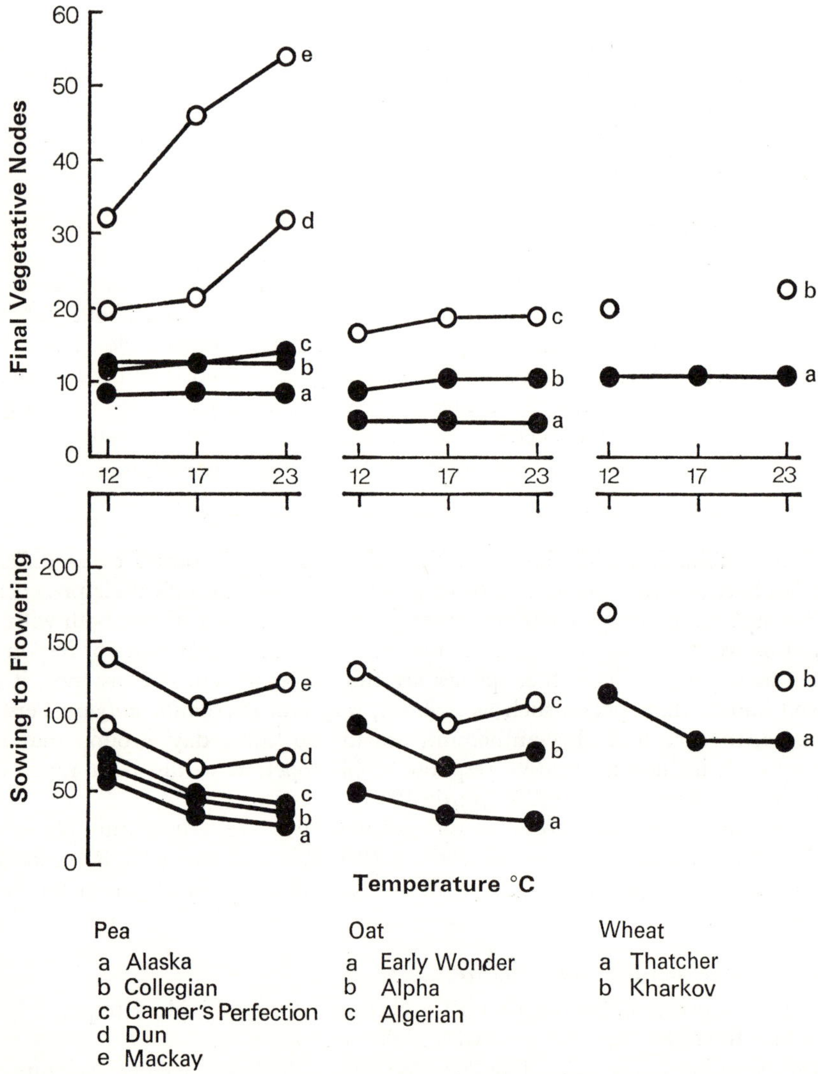

Fig. 5.4 The MG of the variety interacts with the mean daily temperature of the growing season to affect the rate of development. This is shown in the lower section of the figure by the number of days from sowing to flowering, and in the upper section by the VN in early (solid symbol) and later (open symbol) varieties of six species. The values are from the altitude sites 1 cba in the Hawaiian Is at the three mean temperatures of 12°, 17° and 23° respectively. With increasing temperature the variety

creased temperature. VN is unaffected but the period from sowing to flowering decreases. The two later varieties each contain a moderate dose of R3. They therefore reach FI more slowly with higher temperature than when only R1 is involved; VN increases and the period to flowering changes little.

In subterranean clover, even the early variety includes a moderate R3, and the late variety has a much stronger one. The early variety develops fast at 12°C, as shown by the low VN on both main and lateral shoot, because the negative response only slightly increases the number of days to flowering. At 17° the negative response operates more strongly, retarding

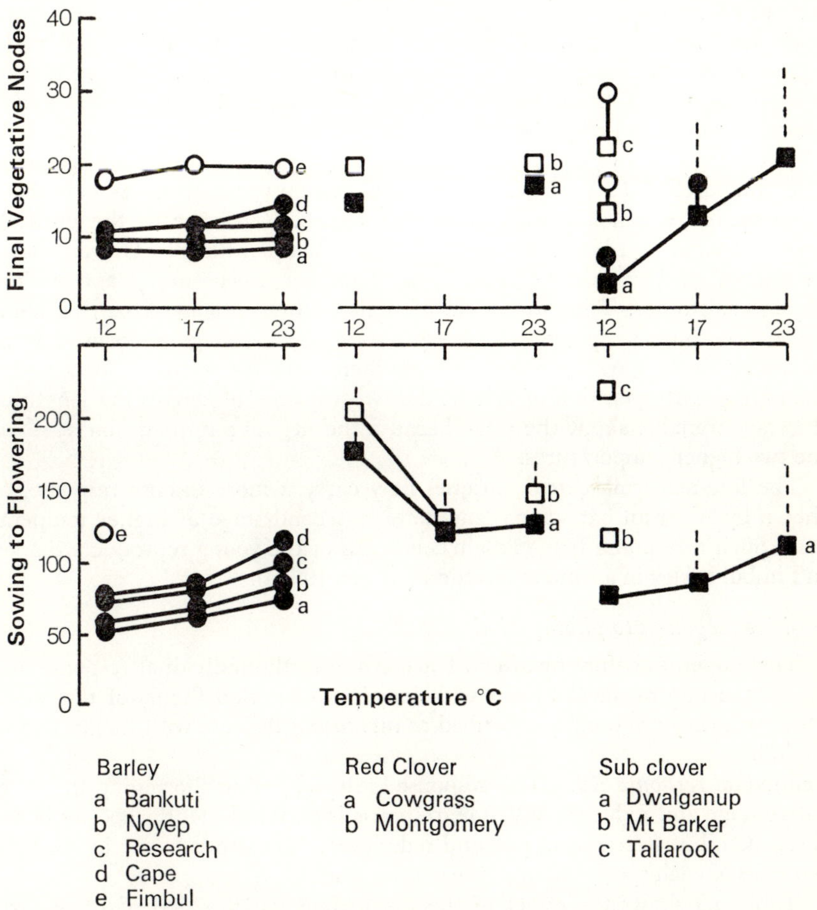

with only universal R1 takes fewer days to flower but has a stable VN. In other varieties, additional responses retard development as shown by the longer period to first flower and by the higher VN. Each symbol (main shoot = circle; lateral = square) indicates the mean of ten plants, and the values for VN usually have a coefficient of variation of less than 2 per cent. Erratic flowering is shown by a dashed line.

FI as shown by the greatly increased VN (from 3 to 12 on lateral). The retardation is only enough to almost counterbalance the effect of R1, thus keeping the length of the period about the same as at 12°. At 23° the negative response is stronger still. The VN is almost doubled (from 12 to 20 on lateral) and much more erratic, and the period from sowing to flowering is lengthened considerably over that at 12° (from about 80 days to over 120).

The effect of a very strong R3 can be seen in the development of the late variety Tallarook, which flowers only at the lowest temperature. Even at this level, FI is delayed as shown by the high VN (22 on lateral). The long period to flowering (over 200 days) also shows that negative R3 is operating in addition to R1. At the two higher temperatures R3 was strong enough to prevent FI entirely.

Individual response R4. The second negative response affects the early reproductive stage (stage 2) in warm to hot growing seasons. Both the rate of leaf appearance of the fixed number of young vegetative phytomers determined at FI (p. 39) and also the rate of floral development may be retarded. Thus the stage from FI to first flower may lengthen with increasing temperature compared with varieties without the response. Only the cereals show this response which is illustrated in Figure 5.6. Reference to the growing period of early and late varieties at the three temperatures in Hawaii shows that three early varieties (Alaska pea, Thatcher wheat and Cape barley) adjoin each other. Thus the effect of temperature on stage 2 can be compared according to species. In pea (without R4) the stage is reduced from 50 to 30 to 22 days respectively with increasing temperature, because of the unimpeded operation of R1. In the two varieties of cereals the length of the stage remains about the same because the negative response increases at the two higher temperatures.

The late-maturing cereal varieties may carry a more intense response, as shown by Algerian oat which lengthens its second stage at higher temperature. Such a response may explain the death of the young reproductive apex in Fimbul barley in a similar situation (Aitken 1966a).

Positive responses to photoperiod

These operate before or after FI according to the individual response and retard the rate of development with longer nyctoperiod (hours of darkness). However, they are usually described as *increasing* the rate with longer photoperiod.

Individual response R5. This response hastens FI, thus shortening the vegetative stage and reducing VN. The response is present in all the species under discussion, but strongest in pea and red clover. It is also stronger in the later varieties of each species than in the earliest ones.

Table 5.3 shows the effect of the mean daily photoperiod of stage 1 on several aspects of development in late varieties of three species. The figures for the late pea, Mackay, are in two sets at different temperatures. Those for each variety are selected from growing seasons matched as closely as possible for the mean daily temperature from sowing to flowering so that differences in rate of development would indicate the effect of photoperiod on R5.

Table 5.3
In late varieties of three species grown at similar temperatures, longer photoperiod reduces the number of days to FI (stage 1) and to flowering (stages 1 and 2) and also the vegetative nodes per shoot

Site	Time of sowing	Species	Variety	Stage 1		Stages 1 and 2		
				Photo-period hr/dy	No. of days	No. of days	Vegetative nodes/ shoot	Temper-ature °C/dy
3b	summer	Oat	Algerian	11·8	120	206	20·0*	13·0
7	spring	,,	,,	17·8	30	85	9·8	14·5
3b	summer	Red clover	Montgomery	12·1	250	320	23·7	12·9
1c	winter	,, ,,	,,	13·3	150	206	18·7	12·3
7	spring	,, ,,	,,	18·0	35	90	6·3	14·7
3b	winter	Pea	Mackay	11·4	118	161	39·0*	11·0
1c	winter	,,	,,	12·0	80	139	29·0*	12·0
10a	spring	,,	,,	24	29	55	17·0	14·1
1b	winter	Pea	Mackay	12·8	90	110	40·0*	17·0
3b	spring	,,	,,	15·5	48	72	26·4†	17·0
4b	summer	,,	,,	15·7	52	77	24·7†	17·0

* some dieback of main shoots
† in this group the mean values of vegetative nodes not significant

In each of the four sets the increase from shorter to long photoperiod causes much faster FI and therefore a shorter vegetative stage and a lower VN. The shorter vegetative stage also reduces the total length of sowing to flowering at the longer photoperiod, though not to the same extent.

The first two sets compare varieties with different intensities of response in the same contrasting seasons. The moderate R5 in Algerian oat reduces the vegetative stage to one-fourth (from 120 to 30 days) and halves the VN because the plants sown in spring at site 7 in Wales grow under much longer photoperiod than when sown in summer at site 3b in Melbourne.

The stronger R5 in the late red clover, Montgomery, reduces the vegetative stage in Wales to one-seventh (from 250 to 35 days) and the VN to one-fourth (from 24 to 6) of the values produced in the plants sown at Melbourne.

The second two sets show that photoperiod affects R5 in Mackay pea in both cool and warm seasons. In contrast to the late subterranean clover, higher temperature does not counteract the trend to a shorter first stage of development, lower VN and shorter time from sowing to flowering with longer photoperiod.

The fact that MG may be based on a range of intensity in R5 is best illustrated by the figures for VN in pea. These varieties have already supplied the evidence for three levels of decreasing tendency to flower (p. 70) which are defined under continuous light at site 10a in Alaska by the minimum VN of 8·5, 12·5 and 17. In addition, maturity in this species is directly related to the presence of stronger doses of R5 as shown by the comparison of the same variety in different photoperiods.

Inspection shows that the VN is stable (not significantly different) for the first early variety, Alaska, in four growing seasons, but increasingly variable in the later varieties. R5 is therefore absent in the Alaska variety but present in slight, moderate or strong doses in the others. For instance the Dun variety was only slightly retarded when grown at site 3b in winter at the shortest photoperiod (11·4 hr/dy), compared with growth at site 10a in continuous light. The moderate response delays development only a little as shown by the few extra VN (+3·5 nodes). In contrast the strong responses in the two late varieties (Mackay and Austrian Winter) at the same sites and times of sowing retard development much more as shown by the greater number of extra VN (+15·5 nodes).

Where the growing seasons are similar in temperature, the length of the vegetative stage is closely related to VN. For instance, when sown in spring at site 3b, compared with site 6, the late pea, Mackay, takes 48 days to reach FI at VN26 in the former case (photoperiod 15·5 hr) but only 35 days at VN19 in the latter (photoperiod 18 hr). The longer photoperiod reduces both measures to about 73 per cent of the values at site 3b.

The same pair of growing seasons provides evidence that the difference between the earliest and the latest varieties lessens at long photoperiod because of the strong response of the later varieties to this factor. At site 3b the earliest variety, Alaska, reaches FI at 6 days from sowing but the late Mackay takes 48 days, a difference of 42 days and 18 units of VN. In the longer photoperiod at site 6 the Alaska variety is slightly slower into FI but without change of VN. Mackay is much faster, taking only another 25 days and 11 units of VN. Thus R5 in Mackay reduces the difference between it and Alaska by 60 per cent in both vegetative stage and VN, compared with the same measures in plants sown in spring at site 3b.

Individual response R6. The second positive response to photoperiod increases the rate of development between FI and heading by faster rates of leaf appearance and of development of the inflorescence. This response occurs only in the cereals and ryegrasses and varies in intensity within both early and later groups. The degree of response to increasing photoperiod also varies, some varieties reaching their maximum at about the 15-hour level while others continue to hasten up to continuous light.

Figure 5.5 shows the development of two varieties of oat in growing seasons differing mainly in photoperiod. The Golden Rain variety has a strong R5 and R6 while the Early Wonder oat has none. In Mexico (site 2) in moderate photoperiod (12 hr/dy) the length of stage 2 of the Golden Rain oat is more than double that of the Early Wonder, because of the retarding action of R6. In Alaska in continuous light (site 10a), the two varieties are alike because stage 2 in the sensitive variety is shortened to the same length as that of Early Wonder.

In the common growing seasons of the temperate zone, photoperiod is too

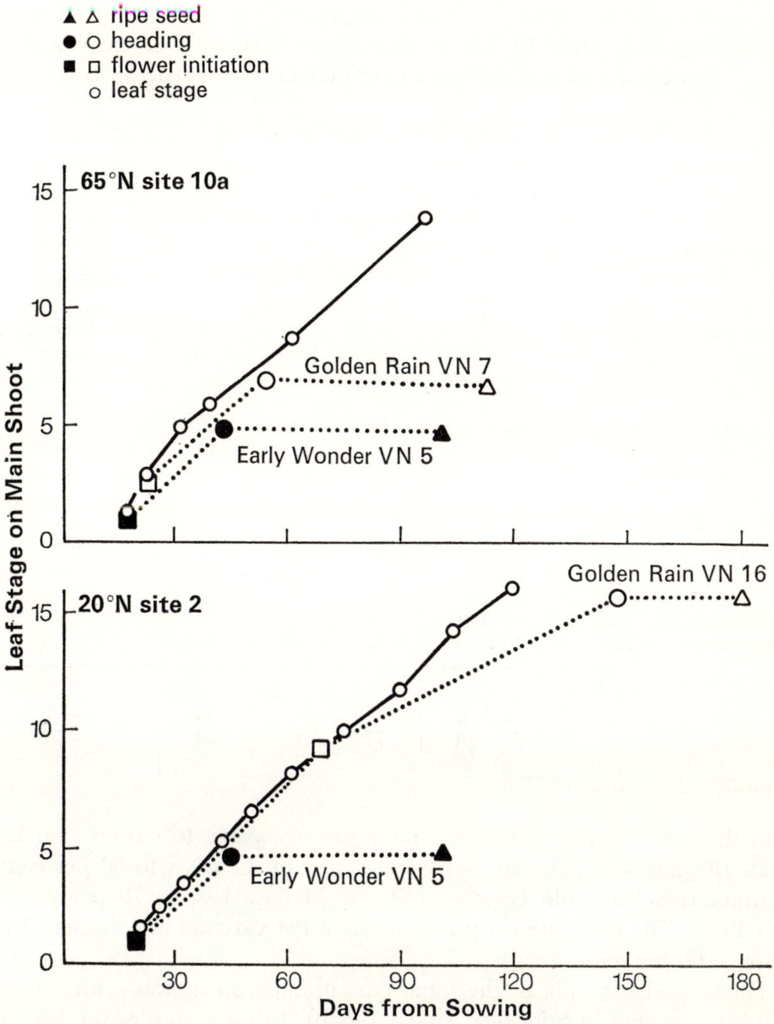

Fig. 5.5 An MG with photoperiod-positive responses develops more slowly in a shorter photoperiod (site 2, Mexico) than in a long one (site 10a, Alaska) as shown in the sensitive oat variety, Golden Rain, compared with the insensitive one, Early Wonder. The gradient of the general slope for leaf stage (continuous line) with days from sowing is similar at both places but the lengths of the first two stages in Golden Rain are much longer in Mexico than in Alaska, and there are many more vegetative nodes.

closely associated with temperature to give useful comparisons. Other examples are therefore derived from experiments with controlled photoperiod at Melbourne (site 3b). In the first experiment, varieties of four cereals were transferred soon after FI to a longer photoperiod than the natural one. Table 5.4 shows that the length of stage 2 in both early and late varieties of each species is shortened in the 15-hour photoperiod and continuous light compared

Table 5.4

The photoperiod after FI affects development stage 2 of early and late varieties of cereals, sown in winter at Melbourne 38°S

Treatment

1. Natural photoperiod after sowing 15 June 1968. Mean hours per day per month from mid-August: 11·55, 12·52, 14·09.
2. Transfer to photoperiod 15 hr about 3 days after double ridge.
3. Transfer to photoperiod 24 hr about 3 days after double ridge.

Species and variety	Maturity	Date of FI	Length of stage 2 (days) Photoperiod			Reduction of stage compared with natural photoperiod (days)
			natural hr/dy	15 hr/dy	24 hr/dy	
Oat						
Alpha	early	15 Aug	68	60	50	−18
Algerian	late	9 Sept	51	42	40	−11
Barley						
Research	early	5 Aug	71	52	52	−19
Fimbul	late	4 Sept	64	51	47	−17
Rye						
Sth. Aust.	early	5 Aug	57	36	36	−21
Varnerag	late	10 Sept	54	42	31	−23
Wheat						
Insignia	early	12 Aug	66		44	−22
Capelle	late	4 Sept	77	62	53	−24

with the natural one. The reduction varies according to variety and is not related to lateness. Algerian oat is shortened by 11 days, to 80 per cent, in continuous light, while Capelle wheat is 24 days less, to 70 per cent (see also Plate 3B). Both the early and the late rye varieties are shortened even more, to 60 per cent.

The second example of the same type of experiment was with ryegrasses and is given here in brief (see Aitken 1966b). In these species the beginning of FI is hard to determine because individuals of the one variety may vary greatly. Clonal material is therefore used for treatments which include longer photoperiod after FI in late winter or early spring. As with the cereals, stage 2 is shortened in longer photoperiods by 30 to 50 per cent compared with the natural photoperiod. The responses to the longer photoperiods (both 14 hours and continuous light) are similar in the three ryegrasses although one of them (Wimmera ryegrass) is an earlier-flowering type than the others.

OTHER INDIVIDUAL RESPONSES AND DEVELOPMENT RATE

Two other responses have been observed in the species under discussion and both modify one or another of the universal responses to temperature.

Individual response R7

This response affects the rate of leaf appearance at a specific level and thus modifies universal R1. Some species have a much faster rate than others at the same temperature (Fig. 6.2 and Table 6.2) as shown by the differences in number of days from sowing to a given leaf stage.

Pea has a very strong response compared with the other species. As shown in Figure 3.11d, the value for days per leaf for pea at 15°C compared with wheat is about 4 and 8 respectively—that is, the pea responds twice as strongly as the wheat. As shown in Figure 5.3b, the number of days from sowing to LS3 is less for pea than for barley and subterranean clover at each temperature. Hence the curve for the days from sowing is lower than for the other species.

Because of the faster response, the matching of a pea variety with another variety of a slower species but the same VN will show a difference in the length of the period from sowing to flowering. A comparison of pea and oat in Figure 5.4 at 12°C shows that the Alaska pea variety (a) and Alpha oat variety (b) both flower at VN9 but the period from sowing to flowering is 60 and 90 days respectively—that is, the pea is 50 per cent faster.

Individual response R8

This response modifies universal R2 by counterbalancing to some degree the retarding effect of low temperature on the rate of seed maturation.

Table 5.5

The effect of seed size, in wheat and pea, on rate of seed maturation and on viability in a growing season changing from warm to cool at site 10a, College, Alaska (mean temperatures of successive periods of 30 days after sowing 29 May 1970 were 12·5, 15·7, 13·6, 6·3°C)

Species and Variety	Developmental measures			Seed measures	
	Stages 1 plus 2 (days)	Stage 3 (days)	VN (number)	size g/100	germination %
Wheat					
Kogat (Alaska)	46	67	6·2	2·4	98
Gabo (Australia)	47	*	6·1	4·1	0
Pea					
—first early					
Alaska	38	66	8·6	26·3	80
Mansholts	43	*	8·9	38·5	0
—second early					
MU38	42	62	12·4	15·4	70
Collegian	45	67	12·2	26·3	60
MU722	51	57	16·0	5·5	55
Canner's Perfection†	51	*	13.0	25·0	0
Frosty†	52	*	11·4	24·4	0

* seed killed by heavy frosts 8-10 Sept
† wrinkled seed type (garden pea)

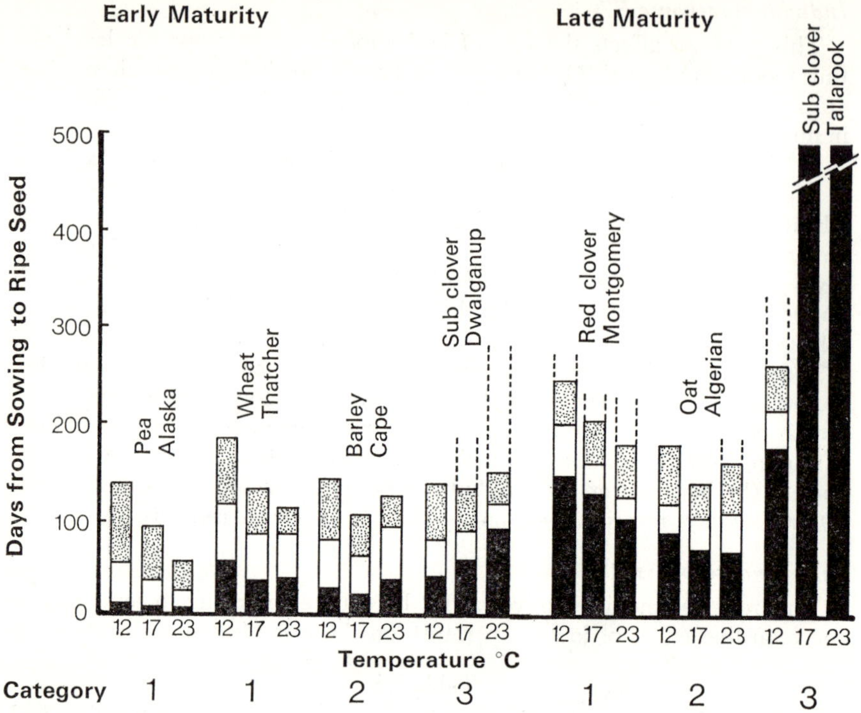

Fig. 5.6 In a tropical latitude at three altitudes (sites lcba, Hawaiian Is), varieties within each maturity group can be classified into one of three categories according to the trend of the growing period with the temperature of the growing season (12°, 17° and 23° respectively). The varieties at the left of each maturity group are in category 1 because the growing period shortens with increased temperature; those at the right are in category 3 because of the reverse reaction and those in category 2 have a growing period that scarcely changes. The varieties range in MG from one with positive responses only to temperature (Alaska pea) to the two with strong negative responses in addition (the two varieties of subterranean clover). Each column shows for one variety at one temperature the three stages within the growing period (vegetative, black; early reproductive, white; late reproductive, stipple). Erratic heading or flowering and ripening are shown by dashed projections to the column.

Its presence was crucial to the viability of seeds of wheat and pea in the growing season in Alaska at site 10a. The slower-ripening seeds were killed by the first hard frosts from 8 September but the faster-ripening varieties escaped damage because the seeds were almost fully dry (Table 5.5).

The presence of R8 may be associated with small size because the Alaskan wheat variety Kogat was originally from the small-grained ecological race from Siberia (Pissarev 1924, Zinserling 1925). The smaller-seeded pea varieties were also faster in ripening seed than others of the same flowering time.

Table 5.5 shows that both Kogat and Gabo wheats began to head at the same time. However, by 19 September the former variety had formed full hard grain which was fully viable while the Australian variety took longer to ripen and the seed failed to germinate.

In pea, R8 was not so clearly associated with size of seed. The earliest varieties of pea began to flower within a few days of each other but only a few ripened seed successfully. In the first early group, the Alaska variety had reached the dry seed stage before the frosts but the larger seeds of the Mansholts variety were evidently killed because they were still immature.

In the second early group, three varieties had ripened seed by the time of the first frosts so that viability was only slightly reduced, in contrast to the complete failure of the other two slower varieties. Two of the three successful varieties had much smaller seeds than the two slower ones. The third successful variety, Collegian, had the same size of seed as the varieties that failed. In this case, R8 may be associated with the starchy character of its smooth cotyledon compared with the sugary character of the wrinkled cotyledon in the two garden varieties.

5 Variation in MG through Combination of Units

The preceding sections have described the two properties of the MG. It remains to discuss briefly the different combinations that may be found within one or another of the maturity groups. The combinations may vary according to species, but the differences may not affect the growing period strongly except in special environments, as will be shown in the next four chapters. The combinations are of two main types.

GENOTYPES WITH SIMPLE COMBINATIONS

The simplest form of MG is one of four units. These include the tendency to flower expressed as minimum VN, the two universal responses to temperature (R1, R2) and the individual response affecting rate of leaf appearance (R7). Figure 5.6 shows the expression of one such genotype in Alaska pea. A low VN (8) and a fast R7 result in two short stages to first flower. At 12°C flowering begins at 60 days but the third stage is slow because of the low temperature, taking another 80 days. The growing period is thus 140 days. At 17° and at 23° each stage is reduced by the operation of the universal responses R1 and R2, making a growing period of 90 and 60 days respectively.

A slightly different MG in oat has the same phenotype as Alaska pea. The variety Early Wonder differs from that of Alaska pea by a lower VN (5) and a slower R7. A comparison of the two varieties when grown at 12° (Fig. 5.4, lower section) shows that the oat heads at about the same time as flowering begins in Alaska pea. The length of the first two stages (sowing to first flower or heading) is the same because the combination of low VN and slower R7 in the very early oat compensates for that of the higher VN and faster R7 in the early pea. The third stage takes about the same time in both varieties because of the same universal R2.

These two forms of MG provide early-maturing varieties in a range of environments because of the low VN and the absence of individual responses to temperature and photoperiod. This kind of combination is theoretically possible in later-maturing groups if a slower tendency to flower were combined with only the universal responses instead of with the additional responses usually present.

GENOTYPES WITH MORE COMPLEX COMBINATIONS

More complex kinds of MG are made up of the above four units plus one or more additional responses. A wide variation in pattern of development is possible from various combinations of these responses as suggested by comparing the formulae for the late variety of oat and both early and late varieties of subterranean clover in Figure 5.1.

Addition of negative responses to temperature

The addition of one or both of the negative responses to temperature, compared with the MG with only the four basic units, lengthens the growing period above a particular temperature threshold. For instance, the presence of R3 retards or prevents FI at higher temperatures according to intensity. In subterranean clover this type of combination in the early variety Dwalganup lengthens the growing period by more than doubling the vegetative stage at 23°, compared with the same stage at lower temperatures (Fig. 5.6). A more intense negative response in the late variety Tallarook prolongs the vegetative stage indefinitely at 17° and above, unless grown in continuous light.

The effect of the addition of both R3 and R4 to the basic four units can be exemplified in barley by the Cape variety (Fig. 5.6). A comparison of development at 12° and at 23° shows that the first two stages are each a little longer at the higher temperature. This is due to both delayed FI (indicated also by VN raised from 10 to 14) and to slower development afterwards. The growing period is thus almost as long as that at low temperature in spite of a shorter third stage. Without the negative responses the growing period at 23° could be reduced to about one-third the length at 12°, as in Alaska pea.

Addition of positive responses to photoperiod

In contrast to the negative responses, the addition of one or both of the positive responses to photoperiod increases the rate of development at a given temperature in a long photoperiod compared with a shorter one. The effect of the presence of R5 can be seen in late varieties of pea (Table 5.1) when sown in spring in Canada (site 6) at a photoperiod of 18 hours compared with the same time in Melbourne (site 3b) at a photoperiod of 15 hours. In Mackay pea the FI is hastened and the VN reduced, as also noted in other examples (p. 77). Because the first stage is reduced from 48 to 35 days, the period from sowing to flowering is shorter (reduced from 72 to 55 days) and the growing period is reduced by 30 per cent (from 112 to 83 days).

MG and growing period

The above examples indicate the basis for a wide range of MG. Depending on the environment and the different combination of units, the phenotypes may differ in growing period or even have the same growing period. A comparison of the different varieties in Figure 5.6 shows much variation in growing period at 23° in both maturity groups. In the early group Alaska pea takes only 60 days while Dwalganup subterranean clover takes nearly 200 days. In the late group the range is wider still. Algerian oat has the shortest growing period (160 days) while late Tallarook subterranean clover remains vegetative well after 300 days.

Figure 5.6 also gives an example of varieties with similar growing periods from different genotypes. If Alaska pea and Cape barley are compared at 12°, it can be seen that they have a similar growing period of about 150 days because of compensating differences in lengths of developmental stages. Cape barley has a longer first stage than Alaska pea but a shorter third stage.

VI Maturity Genotype in Growing Seasons Differing in Temperature

In the previous chapter the concept of MG was explained in terms of two properties, the tendency to flower and a group of responses. Some of the combinations of units were also described which produce differences in MG, both within a maturity group and between extremes of the maturity character.

How far will this analysis help to explain the interaction between MG and climate? The problem is discussed in the next four chapters. These show how different combinations of units in the MG interact with various growing seasons. If the same varieties of plants are exposed to growing seasons with different combinations of temperature and photoperiod, the interaction affects the length of one or more developmental stages and therefore the growing period.

The important principle emerges that varietal interaction with growing season can be classified into one of four categories according to the relationship of growing period to temperature. This helps greatly to organize and clarify the more complex expression of MG.

This chapter considers the simplest type of climatic situation in which the MG is expressed in seasons differing only in temperature. It will however be remembered that the *genetic* situation is not simple, since some of the species include varieties with stronger individual responses to temperature than others, as listed in Table 5.2. Varieties with the simplest MG may be expected to shorten their growing period in warm growing seasons compared with cooler ones because of the unimpeded operation of universal R1 and R2. In other varieties the presence in the MG of the individual responses (R3, R4), when retarded by higher temperature, will reduce the importance of the positive responses. In the extreme cases of the late-flowering varieties of perennial ryegrass and subterranean clover, R3 in each MG is strong enough to completely change the pattern. Instead of the trend to shorter growing period with higher temperature, the trend is to increased length because the negative responses dominate the universal responses to temperature.

The following sections provide three examples of growing seasons with temperature as the main variable. The first includes the seasons at low latitude in the Hawaiian Islands (sites 1a-c) with the same moderate photoperiod. The other two examples include seasons from high latitude with long photoperiod in common. One pair of growing seasons is in Canada and Wales at sites 6 and 7; the other pair is in central and northern Alaska at sites 10a and 10b.

1 Temperature as Variable at Moderate Photoperiod

The plots were sown in winter and the plants grew at three temperatures (12°, 17°, 23°). Both temperature and photoperiod change very little in each growing season from winter onwards as shown in Figure 6.1. The contrasting temperatures of the sites, combined with the moderate photoperiod, reveal the importance of temperature to the length of the growing period. However the interaction is more or less complex according to MG within each of the two groups of varieties, early and late.

The relationship of temperature to growing period is examined first across all varieties of each maturity group, irrespective of species. Further analysis gives examples of genetic differences in rate of development and of their use in classifying varieties into categories of response to temperature.

TEMPERATURE AND GROWING PERIOD

Overall, increased temperature reduces the growing period through the two universal responses R1 and R2. This is seen in Table 6.1 in which the early varieties of nine species are compared as a group with those of late varieties at three temperatures. The mean values include four measures of rate of development and one of leaf stage.

In both early and late groups of varieties, three of the measures are shortened with increased temperature, namely the growing period, the stage from flowering or heading to ripe seed (stage 3), and the days from sowing to LS6. However, the other two measures are not, namely sowing to FI (stage 1), and sowing to flowering (stages 1 and 2). The association of stage 3 and sowing to LS6 with the growing period indicates that R1 and R2 contribute through the rates of leaf appearance and seed ripening. The non-association of stage 1 and stages 1 plus 2 indicates the presence of negative responses in some varieties within each maturity group.

When the information is considered more closely, it will be seen that the results follow from the assortment of MG within each maturity group. The mean value of the growing period of each group at each temperature depends on the number of varieties with and without negative responses and on the intensity of these responses. A variety without negative responses would shorten its growing period with increased temperature as in Alaska pea (Fig. 5.6) in which the growing period at 12°C is reduced to 60 and 40

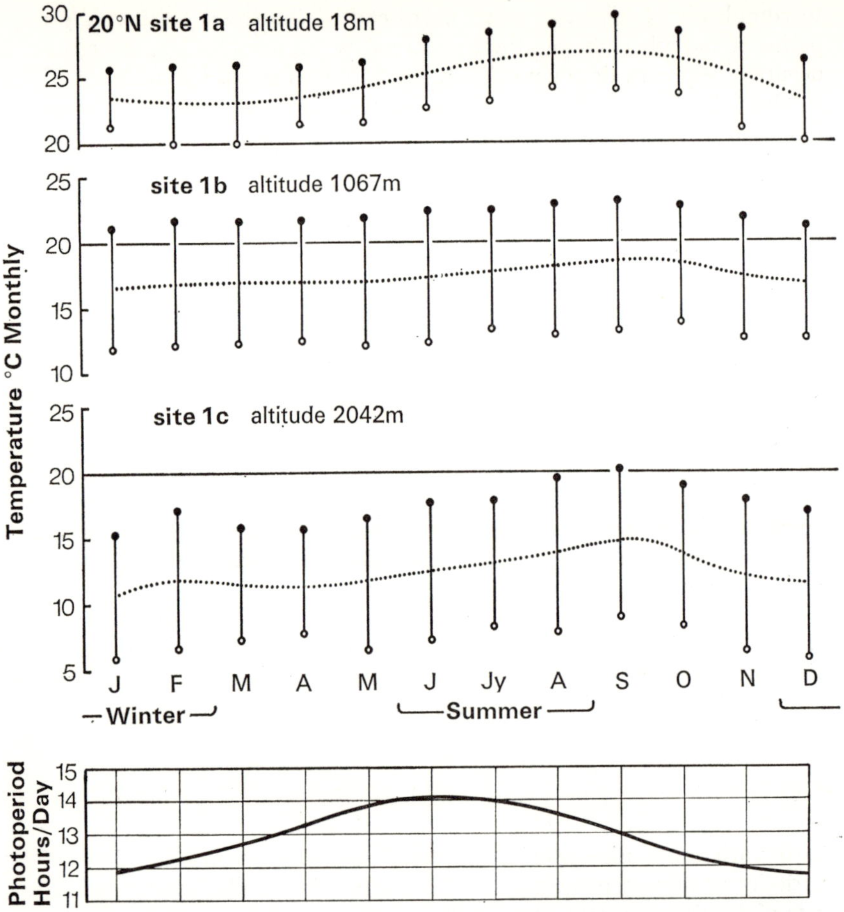

Fig. 6.1 At the same latitude in the tropics, altitude affects the daily temperature but not the photoperiod (including civil twilight) of the growing season, as shown at sites 1a-c in the Hawaiian Is 1963 (see also Fig. 4.9). Sown in winter, the plants at site 1a near sea level grew at temperatures above 20°, but those at higher latitudes grew in cooler environments. The maximum and minimum temperatures (solid and open circles respectively) are placed mid-monthly and the means are connected by the dotted line. The scale is arranged to emphasize the relationship of the temperature at each site to the 20° level.

per cent respectively at 17° and 23°. The fact that the mean value of the growing period is reduced at the warm site and no further at the hot one indicates that some of the varieties have negative responses. For example the early group takes 158 days at 12° but only 65 per cent of this value at both 17° and 23°. Thus the effect of the more complex varieties in the early group is only obvious at high temperature, as shown by the mean growing period remaining the same length instead of decreasing compared with the warm site.

A variety without negative responses would also shorten its first two stages

Table 6.1
The effect of maturity character and temperature on growing period, development stages and leaf stage at three altitudes in Hawaii, as shown by the mean values of all varieties of each maturity group for nine species

Growing season			Site 1c National Park cool	Site 1b Kula warm	Site 1a Waimanalo hot
Maturity group	Varieties no. per site	Measures	Length of each period (days)		
Early	15,14,14	growing period	158	106†	103†
Late	7, 7, 5		226	157*†	165*†
Early	15,15,15	stage 1	37†	28†	37†
Late	10, 8, 6		102†	91*†	96*†
Early	15,15,15	stages 1 plus 2	83†	61†	70†
Late	10, 7, 7		155	109*†	116*†
Early	14,14,12	stage 3	77	51	37
Late	6, 5, 3		72	46*†	41*†
Early	12,13,14	sowing to LS6	44	30†	30†
Late	8, 9, 10		46	31†	31†

* late varieties of perennial ryegrass and subterranean clover omitted because they remained vegetative at the warmer sites
† mean values not significantly different between sites

of development with increased temperature, as in Alaska pea. Because negative responses operate directly over one or both of these stages, their effect is stronger than when considered over the full growing period, because of the compensatory effect of R2 in stage 3. Inspection of the mean values in Table 6.1 for stage 1 of both groups and for stages 1 plus 2 in the early group shows that the lengths of each measure remain about the same with increasing temperature. This indicates that some varieties in each group have strong responses which affect the mean value from 17° onwards.

For example in the early group, stage 1 is 37 days at the cool site and does not change significantly otherwise. This suggests that strong negative responses in the more complex varieties compensate for R1 at both the warmer sites.

INDIVIDUAL RESPONSES AFFECTING DEVELOPMENT TO FLOWERING

Although the universal responses (R1 and R2) shorten the growing period at higher temperatures, this trend may be affected by the addition to the MG of one or another of six individual responses, as defined in the previous chapter. The sites in the Hawaiian Islands have already provided examples of R3, R4 and R7 and further information is presented in this section.

The modifier R7

The presence of positive R7 modifies universal R1, causing a faster rate of leaf appearance in the varieties of one species compared with those of another. Table 6.2 shows that at each temperature the mean value for varieties of each species may differ in the number of days from sowing to LS3. The evidence suggests that R7 varies from very fast (pea) to moderate (cereals) and to slow (subterranean and red clovers).

At the cool site (1c) the mean value of 16·7 days in pea compared with 36 days in subterranean clover indicates that the response in pea is twice as strong as in clover. At the other sites the mean values of both species are lower because of higher temperature, but the difference in rate still occurs.

Table 6.2
The effect of temperature and species on the length of the period from sowing to LS3 at three altitudes in the Hawaiian Islands

Growing season		Site 1c National Park cool	Site 1b Kula warm	Site 1a Waimanalo hot
Species	Varieties	Sowing to LS3 (days)		
Pea	6	16·7	10·0*	10·5*
Oat	3	29·7	17·0	20·0
Barley	5	24·0	15·0	19·4
Rye	2	24·0	16·5	18·0
Wheat	2	28·0	15·0	19·0
Sub clover	3	36·0	25·0	29·0
Red clover	2	36·0	26·0*	26·0*
Ryegrasses annual, perennial	4	33·5	19·8*	20·5*

* mean values not significantly different between sites

Temperature-negative R3 and R4

In contrast to R7 these other responses may annul or reverse the positive trend of the universal responses with temperature.

The R3 response. The presence of R3 in varieties of subterranean clover and pea is illustrated in Figure 6.2 for pea. In this figure the relationship is given, for both pea and oat, between the number of days from sowing and the mean leaf stage. The relationship is shown as a generalized line which slopes upwards from left to right for three temperatures. The time and leaf stage at which FI and first flower occur in an early and late variety of each species is added at the appropriate points on each slope.

Reference to the development of the early pea (without R3) shows that both FI and first flower occur faster at the higher temperature. The latter event is reached at LS9 and takes about 60 days at 12°C but less than 30 days at 23°. The late Mackay pea, however, has an R3 of moderate intensity

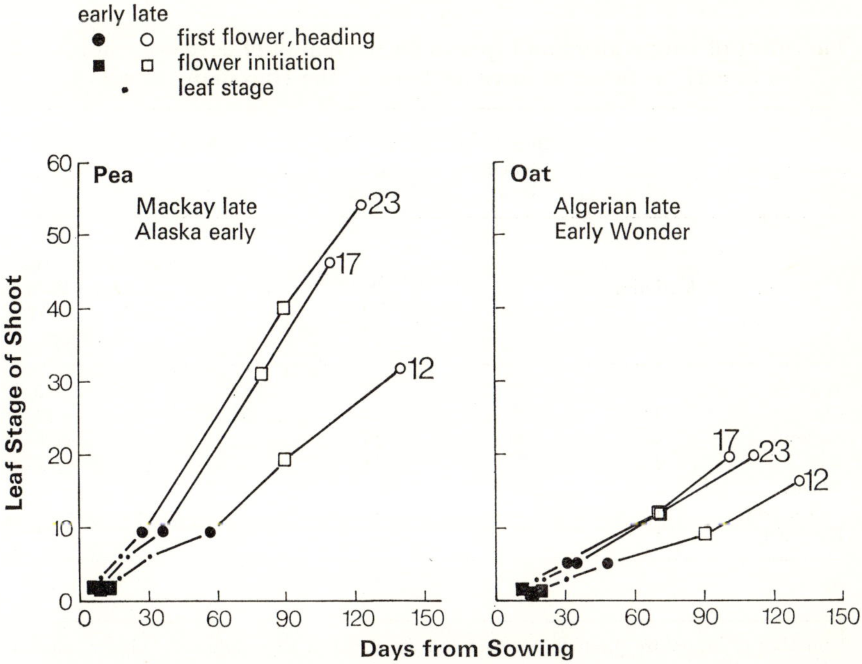

Fig. 6.2 Higher mean daily temperature of the growing season increases the rate of leaf appearance on the shoot, through universal R1, as shown in pea and in oat when grown at three temperatures at sites 1a-c in the Hawaiian Is. The continuous line indicates the generalized slope for leaf stage for the species at a given temperature because early and late varieties are alike until the early one begins to flower. In each species the gradient of the line increases markedly with the rise from 12° to 17°C but not with the next rise. The oat species is genetically slower than pea in rate of leaf appearance, as shown by the gentler gradient of the leaf-stage slope at each temperature.

which retards FI at a higher temperature. Thus the lengths of the vegetative stage at the three Hawaiian sites are similar (90, 80 and 90 days respectively). As a result the VN rises from 30 at 12° to 55 at 23°, as shown by the leaf stage at first flower, and the time of flowering at the three temperatures is much closer than it would have been without R3.

The R4 response. Whereas R3 may affect the vegetative stage of later-flowering varieties of a species more than the early ones, R4 operates in the second stage and only in some varieties of the four cereal species. It slows the development of the few remaining vegetative phytomers after FI, compared with the varieties free of the response.

Table 6.3 gives the length of this stage for early varieties of seven species. The mean values are shortened with increased temperature as shown for a portion of stage 1 (Table 6.2), but the values for the species vary much more at the warmer sites because of weak to strong delay by R4 in the cereals. For instance in ryegrass and subterranean clover, stage 2 is shortened by half at the warmer sites compared with the cool one. The cereals have the same

Table 6.3
The effect of temperature and species on the length of development stage 2 in early varieties at three altitudes in the Hawaiian Islands

Growing season		Site 1c National Park cool	Site 1b Kula warm	Site 1a Waimanalo hot		
Species	Varieties	Length of stage 2 (days)			Loss or gain compared with National Park percentage at K	at W
Pea*	3	40	28	22	−30	−45
Oat	1	60	45	53	−25	−11
Barley	4	48	38	51	−21	+ 6
Rye	1	39	39	38	0	− 2
Wheat	1	60	47	48	−22	−20
Sub clover	1	53	26	23	−51	−57
Ryegrass	1	41	22	20	−46	−51
Mean		49	35	37	−28	−27

* number of immature phytomers above top open leaf at FI significantly higher than in the other species

number of phytomers but the stage varies less. In Thatcher wheat the stage is only shortened about 20 per cent, which indicates a weak response. In barley and rye, stage 2 scarcely changes at the hot site compared with the cool one. This shows the operation of a strong negative response to high temperature which prevents the rate of leaf appearance from increasing.

CATEGORIES OF GROWING PERIOD IN RELATION TO TEMPERATURE

Broadly, the relationship of a variety's growing period to temperature is such as to place that variety in one of three categories (Table 6.4):

1. Growing period shortened by rise of temperature
2. Growing period scarcely affected by rise of temperature
3. Growing period lengthened by rise of temperature

These categories are based on whether the MG contains only universal R1 and R2 or also the one or two negative responses R3 and R4. Whether the growing period of a variety is actually lengthened or shortened with increased temperature depends on the net gain or loss of days resulting from the combination of the positive and negative responses.

Examples of the three categories have been selected from each maturity group to illustrate the contribution of developmental stages to the growing period (Fig. 5.6). The figures for each variety at the three sites in Hawaii are presented as columns arranged in ascending order of temperature. Inspection of the figure from left to right shows four early varieties and three

Table 6.4
The classification of variety into categories of growing period in relation to temperature, plant response, species and maturity group, grown at three altitudes in the Hawaiian Islands

Category	Growing period	Responses	Species	Maturity
1	shortened with rise in temperature from 12 to 23°C	R1 and R2 only or with R3 and R4, slight intensity	pea oat rye wheat red clover	early earliest early early, late early, late
2	little affected	R1 and R2 plus R3 and/or R4, stronger intensities	pea oat barley rye	mid, late early, late early late
3	increased strongly to indefinitely with rise in temperature	R1 and R2 plus R3 and/or R4, much stronger intensities	sub clover ryegrasses barley	early, late early, late late

late ones, all distinct in MG. In the early group the growing period of the first two varieties is shortened with increased temperature, that of the next variety is little affected and the growing period of the fourth one is lengthened with increased temperature. In the late group the same pattern is repeated.

Early group

The first two varieties, Alaska pea and Thatcher wheat, represent category 1 with a reduction in the length of the growing period with increased temperature, associated with no or little change in VN. Comparing the two varieties it can be seen that the reduction in length of growing period between 12° and 23° is larger in the pea than in the wheat (down to 40 per cent compared with 60 per cent). It can also be seen that the main difference between the varieties is in the second stage. In the wheat this stage stays the same length due to the presence of a negative R4 of moderate intensity, while in the pea the stage is steadily reduced with increasing temperature because the expression of R1 is not impeded by R4.

The third variety, Cape barley, represents category 2, in which the growing period changes only a little with increased temperature, because the first two stages are lengthened. This results from the presence of both negative responses (R3 and R4) at moderate levels of intensity. A comparison of the length of the second stage at 23° in both Thatcher wheat and Cape barley suggests the presence in the barley of a stronger R4, judging from the longer stage at the highest temperature.

The fourth variety, Dwalganup subterranean clover, represents category 3, in which the growing period is lengthened by high temperature because the vegetative stage is extended. This is due to the presence of a strong negative R3, which retards FI and thus greatly increases the VN and retards the time to first flower in spite of a faster R1.

Late group

The three late varieties show similar relationships of growing period with temperature. In Montgomery red clover, only a percentage of plants flower at each of the Hawaiian sites, but the flowering ones are graded as category 1 because the growing period is shortened with increased temperature. The VN is unchanged and the positive R1 is able to operate without the obvious complication of negative responses to temperature.

Algerian oat represents category 2, in which the growing period is little affected by increased temperature, mainly because of the two negative responses. R3 retards FI, thus raising VN and lengthening the first stage at high temperature more than in a simple MG. R4 slows the rate of leaf appearance after FI, thus lengthening the second stage in the same manner.

The late Tallarook subterranean clover represents the extreme case in category 3. The shortest growing period is at the lowest temperature. Due to the presence of a very strong R3 in the MG at 17° and above, FI is inhibited, thus prolonging the vegetative period indefinitely.

2 Temperature as Variable at Long Photoperiod

In contrast to the tropical Hawaiian sites, each of the two sets of growing seasons from high latitudes had a longer photoperiod. In addition the tem-

Table 6.5

Temperature and rate of development in stages 1 and 2 in early and late varieties of three species, sown in spring at the same long photoperiod in Canada and Wales 52°N (photoperiod 18 hr/dy in first month after sowing)

Species and variety	Maturity	Site	Length of development stage and temperature				VN number
			Stage 1		Stage 2		
			days	°C	days	°C	
Pea							
Mackay	late	7	36	14·2	30	14·6	17·5
,,	,,	6	36	14·0	20	18·6	19·3
Red clover							
Cowgrass	early	7	30	14·7	60	14·5	5·6
,,	,,	6	34	14·2	27	18·0	6·8
Montgomery	late	7	35	14·2	56	14·5	6·3
,,	,,	6	44	14·3	24	19·9	8·8
Sub clover							
Dwalganup	early	7	15	13·1	45	14·5	3·7
,,	,,	6	27	12·8	25	18·0	5·5

perature during the spring-summer season in Canada (site 6) changed much more in value than in Wales (site 7) or at the Hawaiian sites, while the temperature at Pt Barrow (site 10b) was much colder. Nevertheless the evidence supports the general positive relationship of temperature with rate of development (see p. 87).

TEMPERATURE AND DEVELOPMENT IN SPRING SOWINGS AT 52°N

Temperature and stage 2

Varieties of pea and subterranean and red clovers grown in both Wales and Canada are compared in Table 6.5 for the lengths of their first two developmental stages in relation to the mean daily temperature of each stage. Differences in temperature between the two places are small in the first month but increase thereafter (Fig. 4.9).

Stage 1 is in general little affected but stage 2 is shorter in all four varieties at the Canadian site because the higher temperature in the second month after sowing markedly increases R1 and thus the rate of flower and leaf development.

Temperature and growing period

The early barley, Bankuti, has a fairly simple MG with only a slight R4 and provides evidence of the importance of R2 to the length of stage 3 (flowering to ripe seed) and thus to the growing period (Fig. 6.3). The figure also includes the information from the Hawaiian plots. The major reason for the shorter growing period in Canada compared with Wales (70 and 112 days respectively) is the shortened third stage because of the higher temperature after heading (20° compared with 15°). It can also be seen that the trend of temperature with the length of stage 3 at the sites in Wales and Canada is similar to that occurring at the three Hawaiian sites.

A comparison of stage 2 at the five sites shows that the length is shortened between 12° and 17° but lengthened above this temperature. The reduction in length up to 17° expresses the action of R1 in the remaining phytomers after FI. The increased length at 23° indicates the action of a retarding R4.

The fact that stage 1 takes about 15 days over a range of mean temperatures from 11° to 23° reveals a more complex situation. The stages at both 17° and 23° are alike because FI is slightly retarded at the higher temperature. In contrast the stages with mean temperatures of 11°, 12° and 13° should be longer than that of the stage at 17° because only R1 is present. The evidence thus indicates a higher temperature at plant level in the first weeks after sowing at these sites. This has been found at other sites (Fig. 4.6) with high radiation at sowing time.

TEMPERATURE AS VARIABLE IN ALASKAN SUMMER

The two sites in Alaska may be compared over the short period of 52 days, the length of time from sowing to the end of the growing season at Pt Barrow. During this time both places had continuous light but the mean temperature at site 10b at Pt Barrow was only 2·5° compared with 13·6°

at site 10a at College. At both places the temperature close to the ground was a little higher, but the contrast was still large as reflected in the development of the plants (Table 6.6, Plate 3A). Only five of the earliest varieties were grown at Pt Barrow and these are compared with the same ones at College.

Table 6.6
The effect of temperature on leaf stage and development in early varieties of five species grown in continuous light in Alaska at two sites

		Site 10a College		Site 10b Pt Barrow	
Growing season†					
Temperature °C		13·6		2·5	
Photoperiod hr/dy		24		24	
Species	**Variety**	**Measures of development**			
		Leaf stage and development at 52 days from sowing			
Pea	Alaska	14	mid-flowering	5·5	small flower buds
Barley	Finlay	7	early flowering	3·5	young spikelet
Rye	Sth Aust.	7	early flowering	3·3	young spikelet
Ryegrass	Merredin	8	first heading	2·8	young spikelet
Sub clover	Dwalganup	10	early flowering	1·5	second flower primordia
		Days for development stages			
		Stage 1	Stage 2	Stage 1	Stage 2
Pea	Alaska	11	27	20	all varieties died
Barley	Finlay	17	30	35	
Rye	Sth Aust.	16	21	40	
Ryegrass	Merredin	32	20	45	
Sub clover	Dwalganup	20	28	40	
		Days for development stage 3 and number of vegetative nodes per shoot (VN)			
		Stage 3	VN	Stage 3	VN
Pea	Alaska	66	8·6	—	9·0
Barley	Finlay	58	6·7	—	7·0
Rye	Sth Aust.	*	7·3	—	7·0
Ryegrass	Merredin	40+	8·4	—	6·0+
Sub clover	Dwalganup	*	5·6	—	4·0

* seed killed by low temperature?
† mean daily values for 52 days after sowing

The fact that both the rate of development and the rate of leaf appearance is much faster at College than at Pt Barrow indicates the operation of R1. Stage 1 is halved in length for four of the varieties. For example the early pea reaches FI 11 days after sowing at College compared with about 20 days at Pt Barrow. At 52 days the same variety reaches mid-flowering and is at LS14 at College while at Pt Barrow it only has very small buds and is at LS5.5.

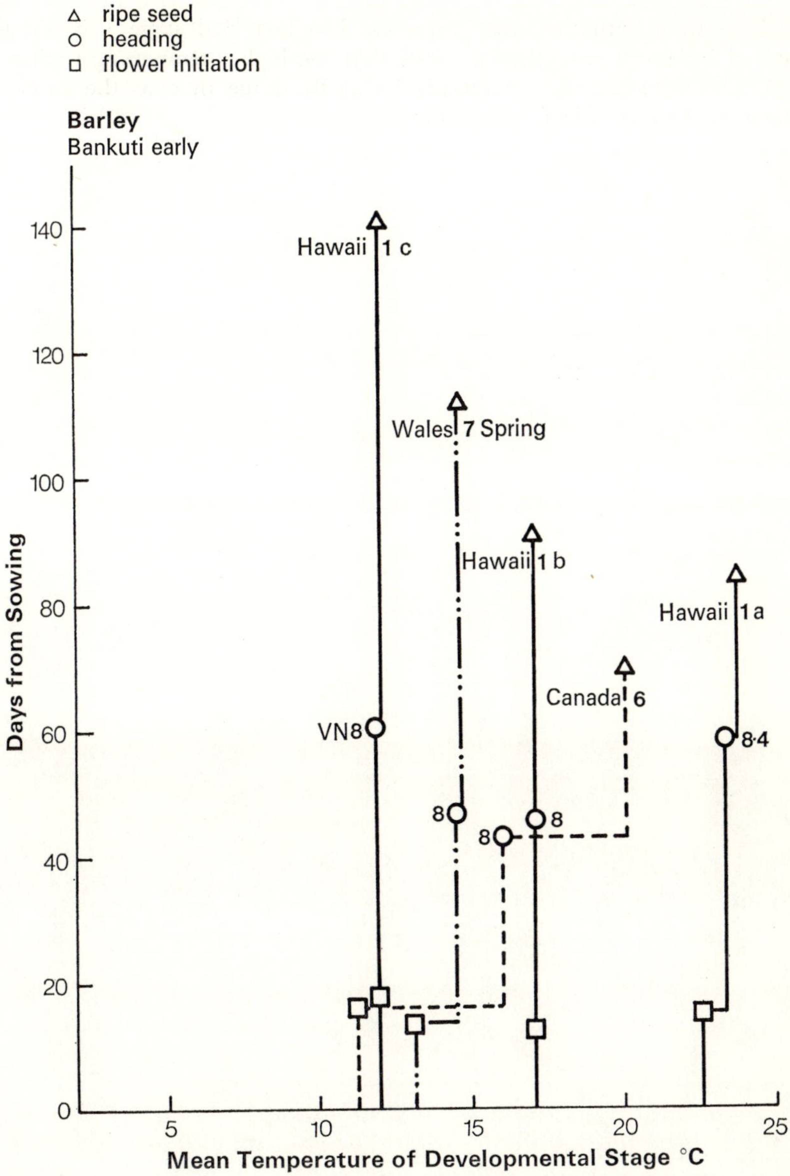

Fig. 6.3 The early Bankuti barley includes the universal responses and also slight temperature-negative ones that affect the first two stages, as illustrated by the relationship of temperature with growing period at five different sites. Each growing period is shown as a line made up of the lengths of the three stages, each placed vertically against the mean daily temperature for that stage. The third stage shortens with increased temperature because of universal R2, but the other two stages change little because of the equalizing effect of the combined positive and negative responses to the temperature.

Only in Merredin ryegrass is the rate of development a little slower at College than with the other varieties. The fact that stage 1 is less than halved in length at College suggests that the higher temperature affects its R3. This response slightly retards FI at the same time as the strong R1 increases the rate of leaf appearance.

VII Maturity Genotype in Growing Seasons Differing in Photoperiod

The discovery that photoperiod could strongly influence the flowering time of plants (Tournois 1914, Garner 1920) was an important physiological advance. It was also simple to demonstrate in sensitive plants by artificially controlling the daily hours of light, but tantalizingly difficult to separate from temperature in the field as an ecological phenomenon. For instance, Evans (1939) found that the flowering of the perennial Timothy grass was hastened by a long photoperiod in the glasshouse. However, clones of the same variety, when planted in a wide latitudinal transect in the United States between Georgia 30°N and Alaska 65°N began the next year to flower later in summer the farther north they were—that is, the *longer* the day length at the start of spring growth!

The contradiction could be caused by the late start to the growing season at high latitudes because of the prolonged winter. It can only be resolved by much more precise knowledge of earlier stages of development in the reproductive shoots and of the interaction of plant responses to both photoperiod and temperature.

The presence of sensitivity within the MG is revealed by observing the reactions of plants in growing seasons differing strongly in photoperiod but little in temperature. It has already been established (Tables 5.1, 5.3) that the development of late varieties of oat, red clover and pea is hastened in longer photoperiods compared with shorter ones. However, a more striking comparison is available from the plots at College, Alaska, and Chapingo in Mexico, where the plants began growth in continuous light or in a 12-hour photoperiod respectively. The temperatures were fairly alike at screen level, and the similar slopes of leaf stage with days from sowing in each species indicate that they must have been closer still at plant level (Fig. 5.5 for oat).

Comparison of plant development in Alaska and Mexico showed that all species had varieties with some sensitivity to photoperiod, but some included more intense responses than others. These are considered in three sections.

1 Sensitivity to Photoperiod within Oat Species

Because of the range of MG in the oat species, it can be used to illustrate the importance of variation in response to photoperiod. Sensitivity to photoperiod is discussed firstly in relation to development over the whole growing period in two varieties, one insensitive and the other very sensitive, and secondly in relation to the length of the vegetative stage in a larger number of varieties with more differences in sensitivity.

PHOTOPERIOD AND GROWING PERIOD

Reference to Figure 5.5 shows that the growing period of the Early Wonder variety is 100 days in both Mexico and Alaska because of its insensitivity, while that of the later Golden Rain variety is 175 days in Mexico but only 115 days in Alaska—that is, 60 per cent shorter. The reduction results from the hastening effect of long photoperiod on R5 and R6 in the first two stages combined with the retarding effect of lower temperature on R2 in the third stage.

In the Golden Rain variety the action of R5 is seen in the reduction of the first stage (sowing to FI) from 75 to 20 days and the VN from 16 to 7. R6 reduces the second stage (FI to heading) from 70 to 30 days. Thus the two responses together shorten the days from sowing to heading in Alaska to one-third of that achieved in Mexico. However, the length of the third stage is doubled in Alaska because the seed ripens during cooler weather. Consequently the two varieties differ less in length of growing period at the two sites than in the previous stages of development.

PHOTOPERIOD AND LENGTH OF VEGETATIVE STAGE

Examination of eight varieties grown in Mexico shows that the days from sowing to FI form a well-spaced series from 20 to 95 days (Fig. 7.1) with corresponding differences in leaf stage at FI and at heading (VN) as shown by the vertical positions of each pair of symbols. These characteristics could be used to classify the varieties into early and late groups. The early group would include the varieties Early Wonder to Nips and the later group, Golden Rain, Algerian, Mugga and Peniarth.

In Alaska the range in days to FI is considerably shorter (from 14 to 63) mainly because the later varieties develop faster than in Mexico. Both earlier and later groups vary in sensitivity to photoperiod. Referring first to the four earlier varieties (Early Wonder to Nips) in Mexico, it can be seen that the vegetative stage ranges from 10 to 55 days. In Alaska, however, the three slower varieties develop as fast as the insensitive Early Wonder, because of the effect of continuous light on the several intensities of R5. Hence the four varieties form one very early group, with a common short vegetative stage of about 15 days. The leaf stage at FI is only about 1·5 and at heading 5 or 6.

The later group includes different intensities of R5. Golden Rain is much

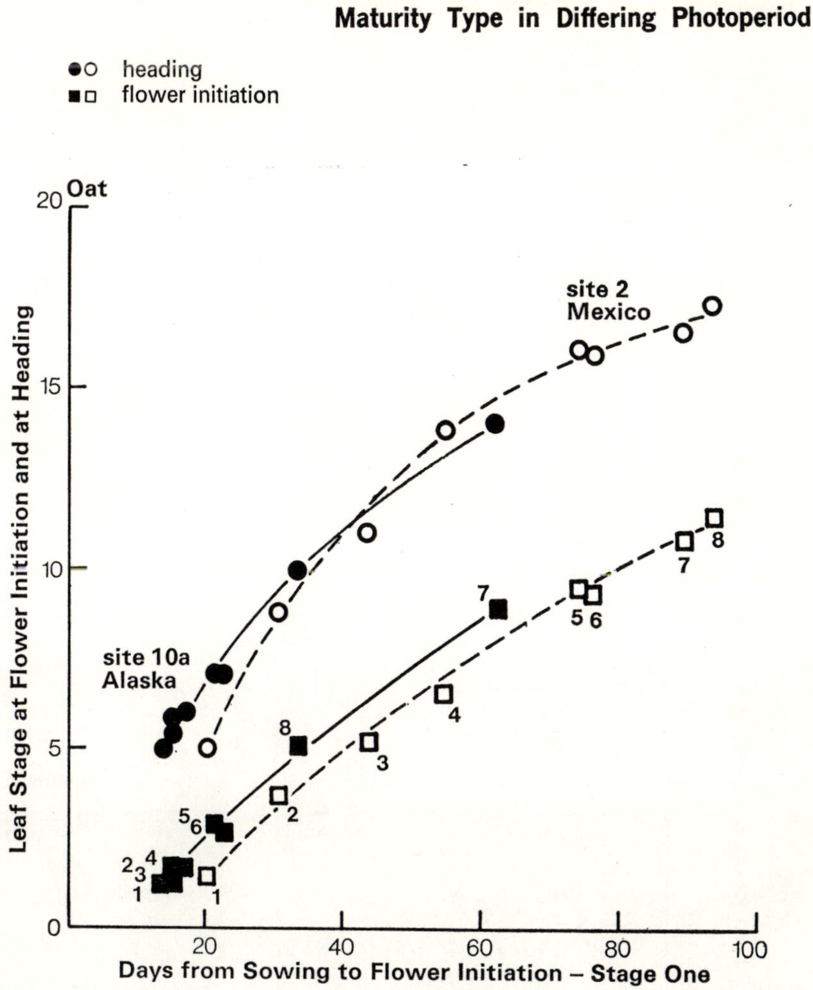

Fig. 7.1 In oat the slower the rate of development in stage 1, the higher the leaf stages at both FI and at heading (VN). This is shown in a range of early and late varieties grown at site 2 in Mexico and site 10a in Alaska. The varieties are numbered in their sequence of flowering in Mexico from earliest to latest: (1) Early Wonder, (2) Rapida, (3) Alpha, (4) Nips, (5) Golden Rain, (6) Algerian, (7) Mugga, (8) Peniarth. In each variety the circle (heading) is directly above the square (FI). Because the later varieties tend to be more sensitive to photoperiod, the range in days from sowing to FI, which is wide in Mexico, is reduced in Alaska, together with the associated measures of leaf stages.

more sensitive than Mugga. The vegetative stage in Golden Rain is reduced to less than one-third (from 75 days in a shorter photoperiod to 23 days in continuous light), while the less sensitive Mugga is only reduced by 30 per cent (90 to 63 days). Thus Mugga remains late-maturing even in the maximum photoperiod. Differences in intensity even transpose the maturity ranking of two varieties. Peniarth is slightly later than Mugga in Mexico but distinctly earlier in Alaska because of its stronger R5.

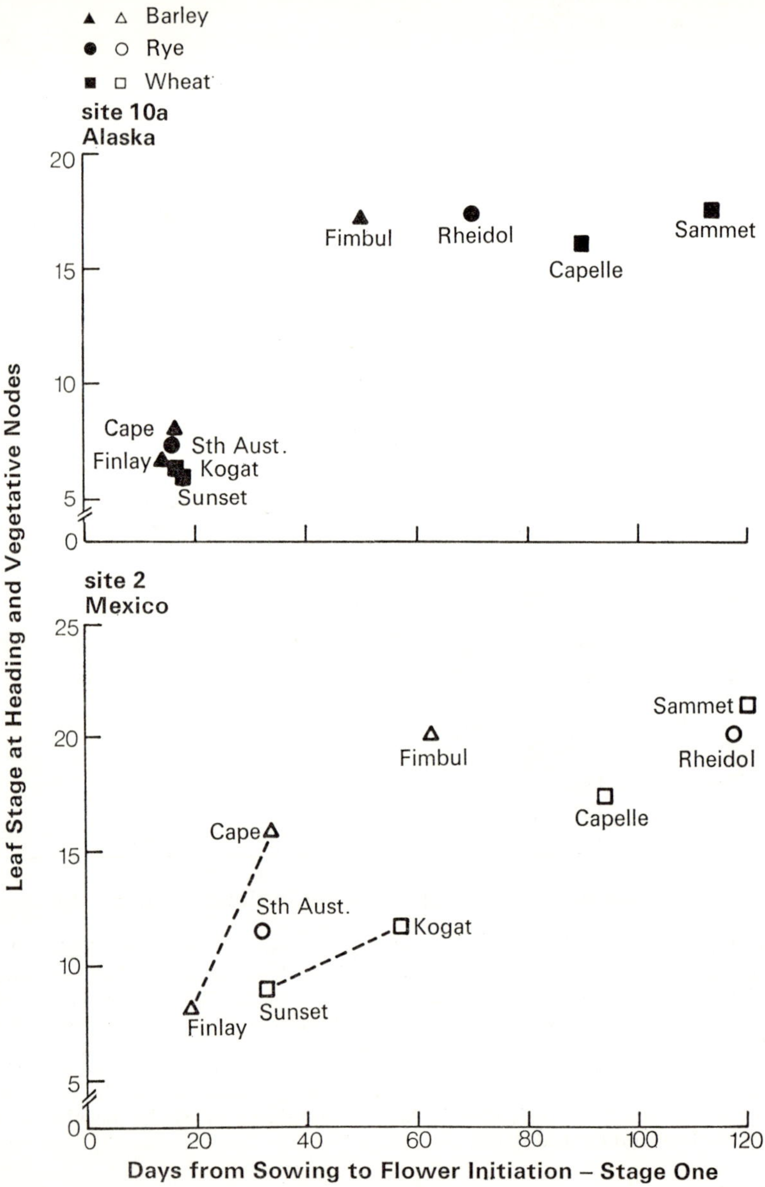

Fig. 7.2 Varieties of cereals vary in sensitivity to photoperiod, as shown by their rate of development to FI and their leaf stage at heading (VN) in a long photoperiod at site 10a in Alaska and in a shorter photoperiod at site 2 in Mexico. In Mexico the width of variation in the early-maturing group is shown by the two pairs of barley and wheat varieties connected by dashed lines. In Alaska the later variety of each pair becomes almost as fast as the much less sensitive one of the pair. In contrast the much later varieties develop slowly at both places, as shown by Capelle wheat.

2 Sensitivity to Photoperiod between Species

As in oat, other species include a range in sensitivity to photoperiod, but in some the range is wider than in others.

CEREALS

The patterns of response in barley, wheat and rye are similar to that in oat. In Figure 7.2 the relationship between the days from sowing to FI and the leaf stage at heading (VN) is shown for samples of the early and the late varieties in each species. The three barley varieties are Finlay, Cape and Fimbul. Finlay has only a very slight R5 as shown by the close position of the two symbols for Mexico and Alaska at VN8·1 and VN6·7 respectively. Cape barley is much more sensitive as shown by the reduction from VN15·7 in Mexico to VN8·1 in Alaska which results from the reduction of the vegetative stage to one-half. In contrast the late Fimbul variety, even more than the late Mugga oat (Fig. 7.1) is reduced only a little in VN and in length of vegetative stage.

In wheat the varieties are Sunset, Kogat, Capelle and Sammet. The earliest wheat (Sunset) has a stronger R5 than the earliest barley (Finlay) as shown by the fact that the vegetative stage is 33 days in Mexico but only 18 in Alaska. In the early Kogat wheat the vegetative stage is reduced further to one-third (57 to 17 days). This response can thus be ranked as more intense than that of Cape barley which only halves the vegetative stage. The two late wheats scarcely change in Alaska compared with Mexico. Both remain with a high VN and a long vegetative stage: Capelle with VN16 takes about 90 days while Sammet with VN18 takes about 110 days.

In rye, a strong R5 reduces the vegetative stage of the early variety (South Australian) to half and the late variety (Rheidol) to about 70 per cent.

Figures 7.1 and 7.2 show that R5 varies within both the early and the later groups of varieties, irrespective of species. In the early group the Early Wonder oat is free of the response and Finlay barley is only slightly sensitive. The most sensitive varieties such as Nips oat and Kogat wheat take nearly 60 days to reach FI in Mexico but only 20 days in Alaska.

In the later group the varieties tend to be less sensitive but Rheidol rye is markedly hastened in Alaska although still slow compared with early varieties. Capelle and Sammet wheat, on the other hand, are only slightly sensitive.

PEA

In pea the patterns of response to photoperiod at the two sites are a little different from those in the cereals. There is a general trend to shorter vegetative stage with continuous light in the cereals and in pea varieties (Table 5.1). The trend, however, is much intensified by increasing strengths of R5 in later genotypes. The range in maturity type is therefore dramatically narrowed in continuous light compared with moderate photoperiod. Figure 7.3 shows the relationship of days from sowing to FI with both leaf stage at FI and at first flower (VN) in seven varieties of early to late maturity.

The early varieties include Alaska (1), Collegian (2) and Canner's Perfection (3) and should be inspected first. In Mexico they reach FI in turn at 14,

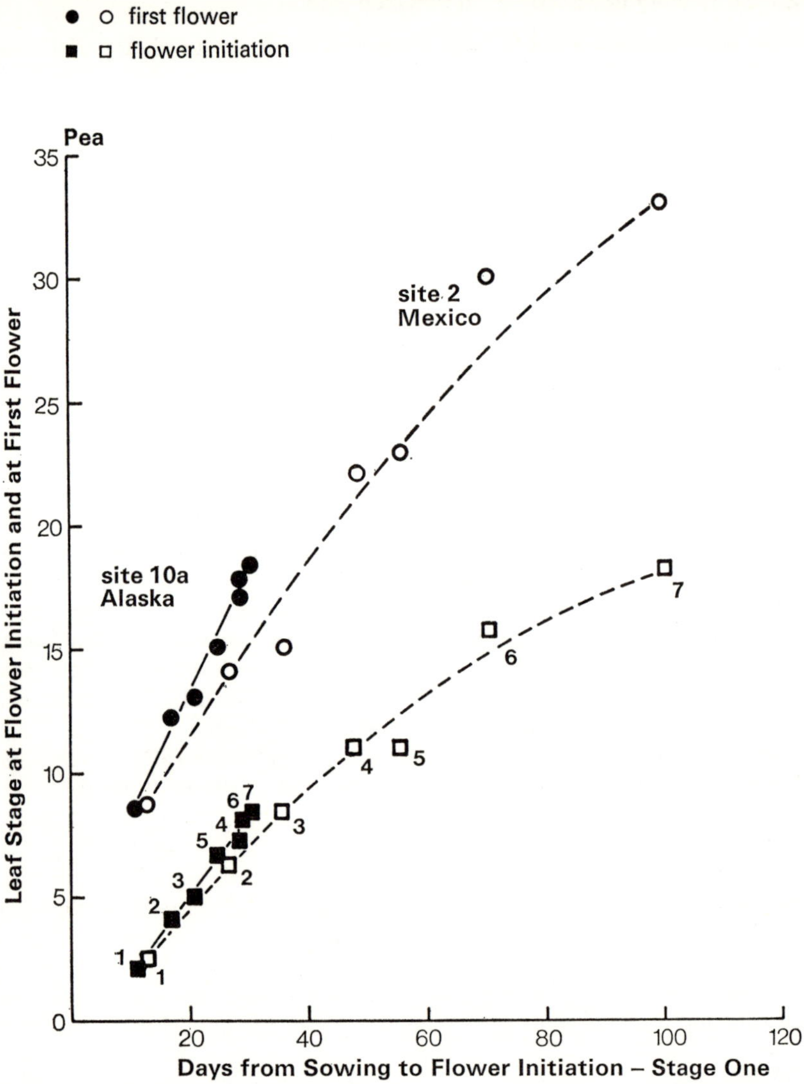

Fig. 7.3 In pea the slower the rate of development in stage 1, the higher the leaf stage at both FI and first flower. These relationships are shown in a range of early and late varieties grown in shorter photoperiod at site 2 in Mexico and in continuous light at site 10a in Alaska. The varieties are numbered in sequence of flowering in Mexico from earliest to latest: (1) Alaska, (2) Collegian, (3) Canner's Perfection, (4) Dun, (5) Partridge, (6) Austrian Winter, (7) Mackay. In each variety the circle (first flower) is directly above the square (FI). Because the later varieties consistently tend to be more sensitive to photoperiod, the wide range in days from sowing to FI in Mexico is greatly reduced in Alaska, together with the associated measures of leaf stages.

27 and 36 days and with a VN of 8·5, 14 and 15. In Alaska the earliest variety is unchanged but the next two pea varieties are faster than in Mexico, reaching FI at 17 and at 21 days and with a VN of 12·5 and 13. This is in contrast with the early oat varieties which contract into the one first early group in Alaska because of a strong R5 in the less early ones.

R5 is more intense in the four later varieties Dun (4), Partridge (5), Mackay (6) and Austrian Winter (7). In Mexico the range in the vegetative stage is between 47 and 100 days, with a VN from about 22 to 33. In Alaska the range is only between 25 and 30 days, with a VN from 15 to 18. The much faster development results from an increasingly strong R5 with lateness of variety. The vegetative stage is halved (reduced from 47 to 25 days) in the late variety Partridge (5) and is reduced to less than one-third (from 100 to 30 days) in the latest variety Mackay (7). As in oat, two of the later varieties change their order of maturity. The more sensitive Partridge is slightly later than Dun to reach FI in Mexico but a few days earlier in Alaska.

In moderate photoperiod the increase in R5 from zero in the earliest variety to very strong in the latest variety results in a wide range in the length of the vegetative stage. In Mexico the range is from 12 to 100 days. In continuous light, however, the latest varieties are hastened most. Thus in Alaska the vegetative stage in all varieties only ranges between 12 and 30 days.

The fact that the late pea varieties develop so much faster in Alaska because of their strong R5 indicates the danger of using 'number of days' as a criterion for maturity type in growing seasons different from that in which the measurement was made. For instance if maturity type in Alaska were judged on the number of days typical of the vegetative stage (up to 36 days) in early varieties in the Mexican growing season, even the latest genotypes would be classified as phenotypically within the early group in Alaska!

OTHER SPECIES

Four other species also include R5 in their range of MG. Both early and later varieties have a shorter vegetative stage in Alaska than in Mexico and a smaller VN (Table 7.1). However these species differ from the cereals and pea in that some varieties remain partially vegetative in Mexico (the two red clovers and the late subterranean clover) and the late variety of perennial ryegrass does not flower at all.

Red clover has a strong R5 in both early and late varieties. The intensity is even greater than in the latest pea variety, as shown by the reduction of the vegetative stage in Alaska to only one-fifth of that in Mexico. Development to the reproductive stage is not only more rapid but more complete than in the shorter photoperiod. The difference between the varieties practically disappears as also in the spring sowings in Wales and Canada (Table 6.5). In addition the main shoot flowers in a proportion of the plants instead of only the laterals, and the percentage is much higher in the early variety.

In subterranean clover R5 is much weaker than in red clover in both kinds of MG. The vegetative stage is only reduced to about one-half in Alaska and the early and late varieties differ strongly in both growing seasons because the temperature is high enough to reveal differences in R3.

Table 7.1
The effect of photoperiod on rate of development in stages 1 and 2 in early and late varieties of four species grown at similar temperature at two sites in Alaska and Mexico

		Site 10a Alaska		Site 2 Mexico	
Growing season* Photoperiod hr/dy		long 24 (60 days)		shorter 12·6 (150 days)	
Species and variety	Maturity	Measures of development			
		Days for development stages			
		Stage 1	Stage 2	Stage 1	Stage 2
Red clover					
Cowgrass	early	27+	33+	150+	30+
Montgomery	late	33+	30+	160+	30+
Sub clover					
Dwalganup	early	20	28	50+	50+
Tallarook	late	60+	30+	125+few veg	20
Ryegrass, annual					
Merredin	early	32+	20+	80	23
Wimmera	late	37+	20+	85	26
Ryegrass, perennial					
Kangaroo Valley	early	27+	20+	80	27
Victorian	late	37+	18+	veg	
		Vegetative nodes per shoot			
		main	lateral	main	lateral
Red clover					
Cowgrass	early	11·0 few veg	5·4	veg	13·3
Montgomery	late	12·6 few veg	5·1	veg	17·0 most veg
Sub clover					
Dwalganup	early	5·6	2·2	10·3	5·0
Tallarook	late	19·0 few veg	13·0	veg	27·0+ few veg
Ryegrass, annual					
Merredin	early	8·4	—	16·0	—
Wimmera	late	9·6	—	17·0	—
Ryegrass, perennial					
Kangaroo Valley	early	7·8	—	17·0	—
Victorian	late	9·7 few veg	—	veg	—

* mean value of photoperiod after sowing
veg vegetative plants

In annual ryegrass the moderate R5 in both varieties appears to be similar to that in early subterranean clover. This also applies to the early variety in the perennial species. The later variety of perennial ryegrass, Victorian, becomes reproductive in Alaska because the continuous light keeps its strong R5 latent. However, in Mexico the same variety remains vegetative because the combination of warm winter temperature and moderate photoperiod activates both R3 and R5. This variety becomes reproductive at moderate photoperiod in the field in southern Australia after growth during the cool winter.

3 Sensitivity to Photoperiod and Growing Season

The previous sections have shown that either or both of the responses to photoperiod can strongly affect the development of a variety when compared in two environments differing only in photoperiod. The Victorian variety of perennial ryegrass, for instance, flowers rapidly in continuous light in Alaska but remains vegetative in the moderate photoperiod in Mexico. The two red clovers also flower rapidly in Alaska but are five times slower in Mexico.

The full growing period may also be affected as with Golden Rain oat (Fig. 5.5) in which a strong R5 and R6 combine to lengthen the first two stages in Mexico three-fold and thus lengthen the growing period compared with Alaska.

The question arises of the importance of varietal sensitivity to photoperiod in the two types of growing season. The patterns of development in selected varieties of pea and cereals are compared in Table 7.2.

IN ALASKA

In central Alaska the growing season is short and in 1970 was only 102 days at College, counted from the date of sowing to the first strong frosts. The temperature falls rapidly from late summer. Seeds must be sufficiently mature by the end of summer to avoid frost damage.

Suitable varieties develop fast because of a low minimum VN, a favourable R8 and a lack of retarding individual responses. These responses may be absent from the MG or else unexpressed in continuous light (positive R5 and R6) or at moderate temperature (R3 and R4). The earliest pea variety reaches FI at emergence of the shoot above ground and is without individual responses except probably a fast R8. This MG permits the seeds to ripen in the available time, but the late varieties, although three times as fast to flower as in Mexico, are too slow.

Some of the early cereals also succeed, by combining rapid FI and heading with a fast R8 as shown in Table 7.2 by the example of Kogat wheat. In Olli barley very rapid FI allows slightly faster heading than the Cape variety, and ripe seed by 84 days from sowing, whereas Cape is a little too slow. Compared with Mexico the Cape variety developed much faster in both stages 1 and 2 in Alaska, yet not fast enough to escape frost damage.

Late varieties such as Winglen wheat are far too slow for the short growing season as shown by their long first stage of 80 days or more and high minimum VN.

IN MEXICO

Turning to Mexico one needs to consider that the high plateau near Mexico City has its wet season in summer. Crops grown in winter must be irrigated, and fast-developing varieties are desirable to reduce the quantities

Table 7.2
The effect of photoperiod on development rate in stages 1, 2 and 3 and on VN in varieties of pea, barley and wheat grown at similar temperature

			Site 10a Alaska		Site 2 Mexico	
Growing season† Photoperiod hr/dy			long 24 (60 days)		shorter 12·6 (150 days)	
Species	Variety	Maturity	Measures of development			
			Days for development stages			
			Stage 1	Stage 2	Stage 1	Stage 2
Pea	Alaska	early	11	27	12	32***
	Mackay	late	29	26	101	40
Barley	Olli	early	15	26	20	56
	Cape	early	17	30	34	79
Wheat	Kogat	early	17	29	57	64
	Gabo	early	21	26	43	55
	Winglen	late	83	**	94	52
			Days for development stage 3 and number of vegetative nodes per shoot (VN)			
			Stage 3	VN	Stage 3	VN
Pea	Alaska	early	66	8·6‡	***	8·5‡
	Mackay	late	*	17·0	43	32·0
Barley	Olli	early	43	7·0	38	7·8
	Cape	early	*	8·6	40	15·7
Wheat	Kogat	early	67	6·2	45	13·2
	Gabo	early	*	6·1	54	9·8
	Winglen	late	—	16·0+	45	17·0

* seed killed by frost during maturation
** young reproductive shoots killed by snow mould in winter
*** buds and flowers killed by frost in winter
† mean value of photoperiod after sowing
‡ mean values of VN not significantly different between sites

of water used and to avoid the poor ripening of seed in the rainy season. However, strong frosts are usual in the winter and tend to reduce the value of rapid development to flowering.

In pea the most useful MG has a moderate R3 and R5. Such a variety is sufficiently retarded by the rising temperature and the intermediate photoperiod to begin flowering after the frosts but before the hot weather.

In the cereals, selection for a short growing period must be based on an MG with a small minimum VN and fairly slight individual responses to photoperiod and temperature. Thus the Mexican barleys are closer to the MG of the Olli variety from Alaska than to the MG of the Cape variety with its strong R5 and R6 which delay the first two stages (Table 7.2). Similarly the modern Mexican wheats are closer to the MG of the Australian Gabo variety than to that of the Alaskan Kogat or the Canadian Thatcher (see Table 10.4). Both of these have a stronger R5 and R6. The late varieties such as Winglen are again inferior for the situation because of a high minimum VN.

VIII Maturity Genotype in Seasons Differing in Temperature and Photoperiod

We have seen how rate of development varies with MG when a variety grows at different temperatures in the same photoperiod (chapter 6), or in different photoperiods at the same temperature (chapter 7). Let us now consider the case when both these climatic factors vary. This is the common situation, especially in temperate climates, although in the tropics both temperature and photoperiod change little from month to month at any one altitude. Hence time of sowing scarcely affects the level of each factor. In a higher latitude the monthly rate of change of both factors is faster and the range between summer and winter is much larger. Thus the time of sowing during the year may strongly affect the type of growing season available for the plant.

1 Patterns of Development in Simple and Complex MG after Sowing through the Year

In a favourable climate the same varieties of crop can successfully complete their growing period even though sown in different seasons of the year. As the climate becomes less favourable in one part of the year, the desirable times for sowing useful crops are greatly reduced and the choice of species becomes more limited.

In southern Australia, apart from the highlands, the climate is favourable to growth all the year provided that sufficient water is available. Temperate species grow slowly in the cool winter and faster in the warm seasons whether the temperature is rising as in spring or falling as from late summer.

The figures from a series of sowings through the year at Melbourne (site 3b) show the developmental pattern of a simple MG in oat—that is, one with only the universal responses to temperature in changing values of temperature and photoperiod (see p. 58). The pattern should be compared with those of more complex MG in Figures 8.1–8.4.

Figure 8.1 shows the relationship between time of sowing and the number of days to FI and to flowering in early and late varieties of oat and subterranean clover. In addition the length of the growing period is shown for the two early varieties at three of the sowing times.

Reference to Figure 4.7 gives the seasonal changes in temperature and photoperiod. Beginning from the first month of winter the temperature remains about the same for the first three months and the photoperiod rises only slowly. Plants sown in early spring grow in rising temperature and photoperiod while those sown in mid-summer grow in falling values which change more rapidly in photoperiod than in temperature.

AN EARLY VARIETY OF SIMPLE MG

The pattern of development in Early Wonder oat is one in which the longest growing period (180 days) follows sowing in early winter mainly because the ensuing low temperature slows both leaf appearance and inflorescence development through R1. Conversely plants sown both in spring and in late summer develop more rapidly and ripen seed after only 90 days because the higher temperature hastens both R1 and R2.

Figure 8.1a shows that the length of stage 1 (sowing to FI) varies from about 10 to 40 days. The lengths from successive sowings form a curve that is at its maximum after sowings in winter and at its lowest after sowings over summer.

The length of stage 2 (FI to heading) varies in a similar way (30 to 90 days) but is longer than stage 1 at each sowing. This occurs because of the constant low VN5. Whereas only LS1·3 must be reached before the shoot apex becomes reproductive, nearly 4 more leaves must expand fully on the shoot before the plant heads.

Comparison of the lengths of the three stages after sowings in spring (October) and summer (February) shows that complementary differences in the length of stages 2 and 3 (heading to ripe seed) result in similar growing periods. Because of the seasonal curve of temperature, plants sown in spring pass through stages 2 and 3 while temperatures are rising; hence the latter stage is the shorter. Those sown in late summer pass through the same stages while temperatures are falling; hence the latter stage is the longer. The sudden increase in the length of stage 2 that follows sowing in early autumn (March), results from the falling temperatures through late autumn and early winter (April, May and June).

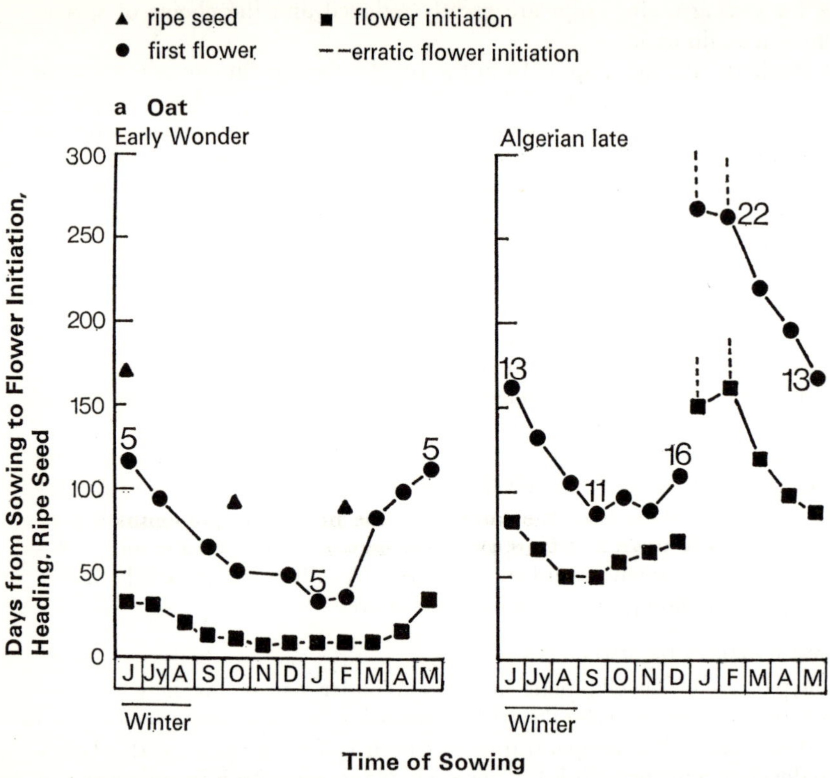

Fig. 8.1 The MG of a variety may be revealed by the way the time of sowing through the year affects rate of development to FI and to heading or first flower, as shown in an early and a late variety of oat and subterranean clover at site 3b in Melbourne. The early oat has the simplest pattern of development, with the growing period

With the above pattern in mind the MG with extra responses may be better understood.

AN EARLY VARIETY OF COMPLEX MG

Reference to Figure 8.1b shows the different pattern of the early Dwalganup variety of subterranean clover. The longest growing period follows sowing in late summer. The curves for stages 1 and 2 are both opposite in timing to those for Early Wonder oat, with the shortest lengths from the plants sown from mid-winter to spring and the longest from those sown in late summer. The VN also changes from the minimum of 5 on the main shoot in plants sown in winter to an erratic level of 15 or more when sown in summer.

The pattern marks the presence of a strong R3 which begins to delay FI increasingly from the start of sowings in spring as the temperatures rise. The maximum delay follows the late summer sowing because FI is only reached erratically by late autumn, after which flowering and seed ripening are delayed greatly by the low temperature of winter.

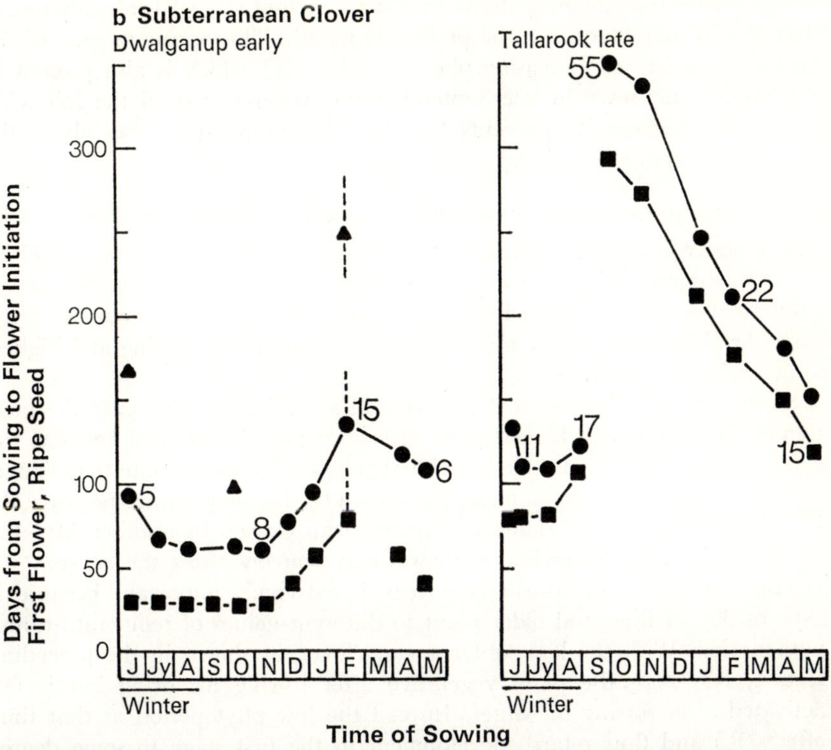

shortest at high temperature and longest at low temperature. The other varieties have more complex patterns. The number of vegetative nodes on the main shoot is shown for selected times of sowing and also the number of days to ripe seed.

TWO LATE VARIETIES OF COMPLEX MG

A comparison of Algerian oat, a late-maturing variety, with the early variety shows a striking difference. Both the first and second stages are at their longest after mid-summer sowings and at their shortest after spring sowings. In addition the minimum length of stage 1 is 50 days, which is much longer than that of the early variety at the same time of sowing (15 days).

The pattern is caused mainly by a strong R5 together with a moderate R3. A comparison of the two sowings in summer (December and January) shows that the December-sown plants took 70 days to reach FI by late February. The January-sown ones, however, remained vegetative until winter (150 days or more) because R3 and R5 inhibited development while warm temperature and lower photoperiod were combined from mid-summer through autumn.

A comparison of the late-maturing Tallarook subterranean clover with the early oat shows still greater differences. The longest period for stage 1 follows sowing in mid-spring and the shortest period follows winter sowings. The pattern results mainly from a very strong R3. Inspection of the sowing in late winter (late August) shows that the length of stage 1 is already longer than that of plants sown in the preceding month. This occurs in spite of the favourable effect of increasing photoperiod on R5 which is also present in the MG. Plants sown in late September stay vegetative until the following late winter because R3 prevents FI when the temperature rises above the threshold level in late spring.

MG AND CHANGING VALUES OF TEMPERATURE AND PHOTOPERIOD

The above examples establish the patterns of development of several kinds of MG as curves related to times of sowing. In the simple MG, rate of development is directly related to temperature only. Hence the curves of the three developmental stages are lowest after sowing in late spring and highest after sowing in winter.

A complex MG may affect the curves in several ways, changing the position of the lowest and highest parts of each curve, the contrast between the extremes, and the shape. One individual response may dominate the change, depending on its type and intensity, as shown by the temperature-negative R3 in the early and late varieties of subterranean clover. In another MG the presence of several individual responses may scarcely affect the curve.

The time of sowing during the year is extremely important because it exposes the seedling and older plant to different values of temperature and photoperiod. With the late subterranean clover, as shown in the preceding section, R3 keeps the plant vegetative after sowing in spring but is not activated after sowing in winter. Instead the low photoperiod at that time affects R5 and thus retards development in the first stage to some degree (Aitken 1955a). In other kinds of MG, time of sowing may be critical to successful setting of seed because of responses in the later stages that may block development, as with late cereals when sown in spring (see next section).

The differing patterns of development of simple and of complex MG (Fig. 8.1) during the year provide the key to an understanding of the expression of MG in different growing seasons. This problem is examined through five groups of growing season, the first concerned with development after three times of sowing at Melbourne, and the others after sowing in different types of locality.

2 MG in Cereals and Time of Sowing at Melbourne

Early-maturing Australian varieties of the temperate cereals are similar in MG. In addition to a faster tendency to flower, they are likely to have smaller responses negative to temperature (R3, R4) and positive to photoperiod (R5, R6) than those in late-maturing varieties. Because of these responses, the early group develops in a different way from the late group after sowings in winter, spring and summer. Figure 8.2 gives the results for one early and one late variety of wheat, barley and rye.

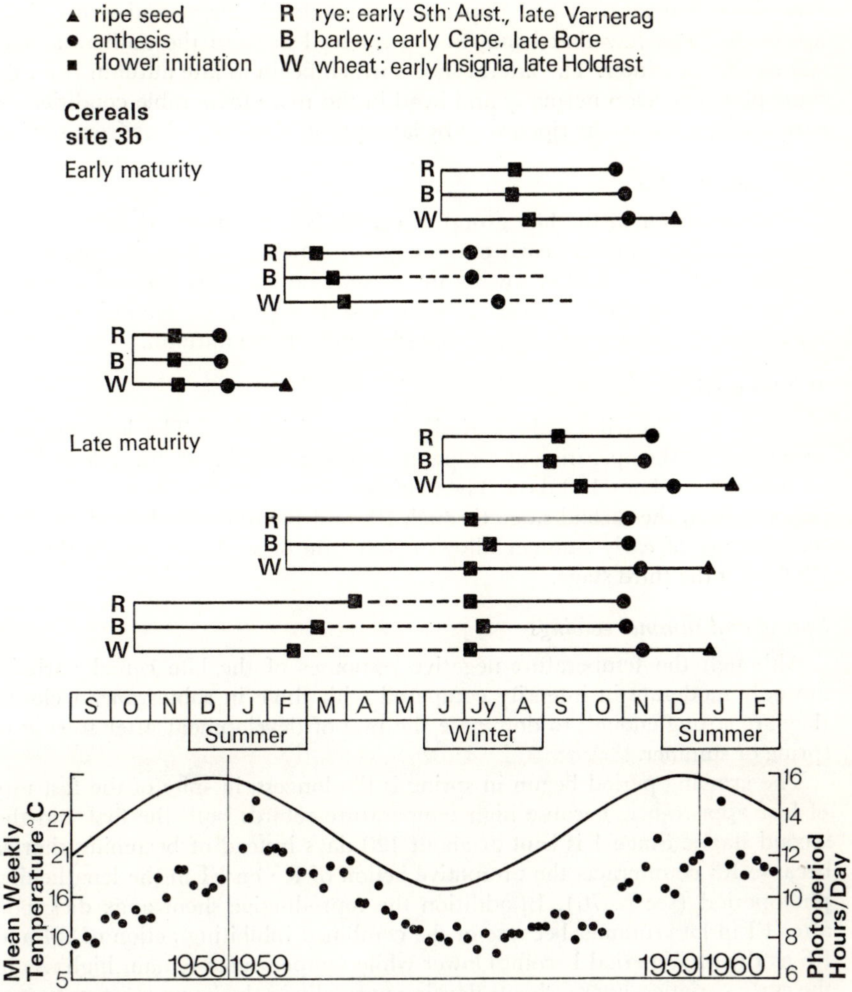

Fig. 8.2 In cereals the time of sowing at site 3b in Melbourne affects the length of the growing period in both early and late varieties. The two groups differ most when sown in spring because of the breakdown in development soon after FI in the late varieties.

EARLY GROUP

The growing period of the early varieties is shortened by almost half after sowing in spring compared with that of winter (e.g. Insignia wheat takes 120 days compared with 200). This behaviour resembles that of the very early oat (Fig. 8.1), which indicates that the universal responses (R1, R2) almost entirely control the rate of development. However, in contrast to the oat, the growing period is longer after sowing in summer than in spring because the small R4 and R6 are sufficient to delay stage 2 into the low temperature of winter. Most shoots reach FI rapidly in early autumn but retarding responses slow down the rates of leaf appearance and development and many shoot apices die. The few that head, fail to set seed because the anthers and/or stigmas die in winter. The lateral shoots which begin in late autumn from the same plants develop normally and head in the more favourable conditions of early spring. These ears ripen seed by late spring, about 170 days from sowing.

LATE GROUP

The behaviour of the late group of cereals is even more striking. In complete contrast to the very early oat, the growing period is shortest after sowing in winter and longest after sowing in spring. The length is doubled as shown by Holdfast wheat which takes 220 and 450 days respectively. This indicates the action of strong retarding responses in summer and autumn.

Winter sowing

When sown in winter, the varieties develop fairly fast. The low temperature prevents the operation of temperature-negative R3 and thus allows FI to be reached in about 120 days. Then rising values of temperature and photoperiod hasten the second stage through R5 and R6 respectively and the high temperature of early summer allows a fast rate of seed maturation through R2 during the third stage.

Spring and summer sowings

Although the temperature-negative responses of the late cereal varieties are only moderate in intensity compared with those in subterranean clover, they are strong enough to dominate the rate of development after sowing in spring or summer.

The growing period begun in spring is the longest, in spite of the fast rate of leaf appearance, because high temperature retards both the first and the second stages. Stage 1 is kept at about 120 days instead of becoming shorter because R3 counteracts the promotive action of R5 on FI in the lengthening photoperiod (see p. 76). In addition the reproductive shoot apex dies soon after FI in late summer because of the combined inhibiting action of R4 and R6 as the photoperiod becomes lower while temperatures remain high. As in the early varieties, lateral shoots develop normally in the lower temperature of late autumn and reach FI in winter, 250 days after the plants are sown. The second stage in the laterals is then retarded to some degree by the lower photoperiod through the action of R6. However, stage 3 is rapid in the higher temperatures of late spring and early summer because of R2, and the growing period ends at 450 days.

Plants sown in late summer follow a similar pattern of development except that the first-formed reproductive apices develop normally after FI in winter.

3 MG in Three Types of Growing Season in Temperate Zone

Three different kinds of growing season within the temperate zone are exemplified in Melbourne at 38°S and in Saskatoon (Canada) and Aberystwyth (Wales) at 52°N. The importance of growing season to the expression

Table 8.1

In four contrasting seasons at three latitudes, maturity character interacts with the temperature and photoperiod to affect the development rate of varieties of pea, wheat and subterranean clover

			Site 3b Melbourne sown winter	Site 6 Canada sown spring	Site 7 Wales sown autumn	Site 10a Alaska sown spring
Growing season†						
Temperature °C for 90 days			9	17	11	14
Photoperiod hr/dy for 30 days			10·6	17·6	12·8	24·0
Maturity	Species	Variety	Measures of development			
			Days for development stage 1			
Early	pea	Alaska	13	10	18	11
	wheat	Gabo	71	28	45	21
	sub clover	Dwalganup	27	27	28	20
Late	pea	Mackay	118	35	*	29
	wheat	Kharkov	124	120?	190	114*
	sub clover	Tallarook	100	280?	160	52+
			Days for development stage 2			
Early	pea	Alaska	67	25	35	27
	wheat	Gabo	56	19	*	26
	sub clover	Dwalganup	90	25	180	28
Late	pea	Mackay	42	20	—	26
	wheat	Kharkov	49	270?	76	*
	sub clover	Tallarook	50	60?	90	30+
			Days for development stage 3			
Early	pea	Alaska	66	33	**	66
	wheat	Gabo	53	33	—	**
	sub clover	Dwalganup	53+	38+	55	**
Late	pea	Mackay	40	28	—	**
	wheat	Kharkov	45	40	60	—
	sub clover	Tallarook	48+	50?	55+	**,*
			Number of vegetative nodes on shoot (VN)			
Early	pea	Alaska	9·1	8·5	8·4	8·6
	wheat	Gabo	9·0	7·8	8·0	6·1
	sub clover	Dwalganup	2·5‡	4·5‡	4·1‡	5·6 2·2‡
Late	pea	Mackay	32·4	17·0	—	17·0
	wheat	Kharkov	17·5	20·0	16·0	17·0
	sub clover	Tallarook	8·0‡	50·0?‡	15·0‡	19·0veg 13·0‡

* vegetative or young reproductive shoots killed by frost
** developing seed killed by frost
† mean values of temperature and photoperiod after sowing
‡ value on lateral shoot
veg vegetative plants 30%
? estimated figures from same variety in less extreme growing seasons on the basis that Kharkov wheat was close to FI 80 days from sowing while Tallarook was vegetative and also survived the following winter

of MG is illustrated with early and late varieties of three species grown at these places (Table 8.1).

The figures from Melbourne give the pattern of development of each variety in southern Australia after sowing in early winter and therefore in low temperature and short photoperiod (see p. 56). Those from Canada give the pattern of development typical of the same variety in the warm spring-summer growing seasons of inland Europe and of North America. Those from Wales represent the plants sown in the cool autumn and harvested the following summer in the maritime areas of these continents.

At Saskatoon, plants sown in late spring grow in rapidly rising temperature and in long photoperiod for the first three months and then in lower values of both factors, particularly temperature, until the following spring (Fig. 4.9). In contrast, those sown at Aberystwyth in early autumn, begin to grow in lowering values of temperature and photoperiod. The temperature remains below 10°C from late autumn to late spring while the photoperiod begins to lengthen from late winter onwards.

In each species, inspection of the developmental stages of each variety at the one site shows their contribution to the growing period and indicates the action of the most important responses.

EARLY GROUP

All three varieties develop most rapidly at Saskatoon (site 6) because of higher temperature, and the differences in growing period between this and the other two places are entirely or largely due to universal R1 and R2, according to MG.

In Alaska pea the simple MG shortens the growing period at Saskatoon to about half that at Melbourne (from 146 to 68 days) by doubling the rates of leaf appearance and of seed ripening. At Aberystwyth the variety reaches first flower a little faster than at Melbourne because of slightly warmer conditions, but dies soon after as the autumn cold begins.

Gabo wheat has an MG somewhat like that of the Insignia variety (p. 115). The growing period is more than halved at Saskatoon compared with Melbourne (from 180 to 80 days) mainly because universal R1 combines with individual R5 and R6 to shorten stages 1 and 2 (Fig. 8.3). In the longer photoperiod at Saskatoon, R5 hastens the rate of development by early FI (as shown by lower VN), and R6 increases the rate of leaf appearance in stage 2.

However, when sown in autumn at Aberystwyth, the plants fail soon after FI. They reach this stage a little faster than at Melbourne (by 45 days instead of 71) but at the same VN. As the photoperiod shortens in late autumn, R6 retards the reproductive apex and this is soon killed by frost.

In Dwalganup subterranean clover the MG includes a strong R3 (p. 112). This retards FI slightly at Saskatoon and Aberystwyth, compared with the cooler Melbourne temperature. The VN on lateral shoots is raised from 2·5 to 4·5 but FI occurs within the first month after sowing at the three places because of the compensating effect of R1. This counterbalances the raised VN by faster rate of leaf appearance at the two warmer places.

Differences in growing period between the sites result almost entirely from

Maturity Type in Differing Temperature and Photoperiod

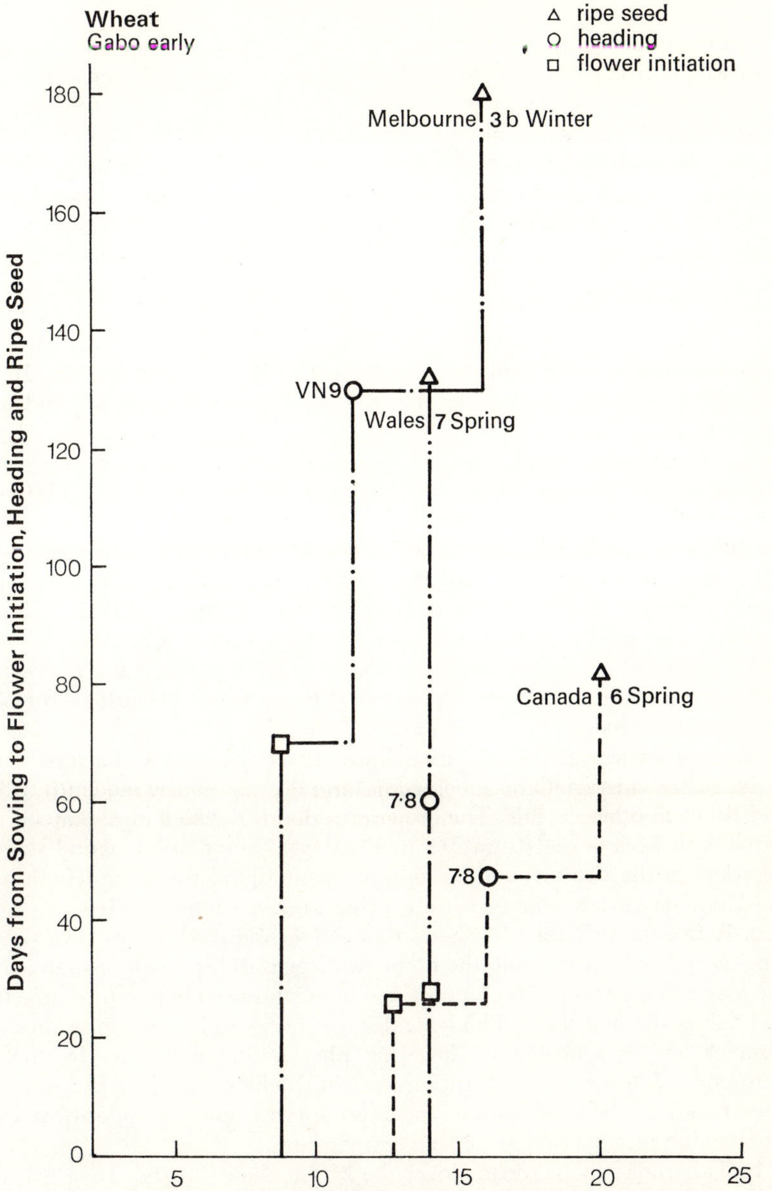

Fig. 8.3 In early Gabo wheat, both the universal responses and a slight sensitivity to photoperiod may affect the length of the growing period, as shown by winter-sown plants at site 3b in Melbourne compared with those sown in spring at sites 7 and 6 in Wales and Canada respectively. Each growing period is shown as a line made up of the lengths of the three stages, each placed vertically above its mean daily temperature. The trend to shorter growing period with increased temperature results not only from the universal responses shortening all three stages but also from the more rapid development in the higher latitude. One measure of the hastening effect of longer photoperiod is seen in the lower VN, reduced from 9 to 7·8.

R1 in stage 2. The growing period is almost halved at Saskatoon compared with Melbourne (from 170 to 90 days), because of much faster leaf appearance after FI in the higher temperature. At Aberystwyth the growing period is nearly doubled compared with Melbourne (from 170 to 263 days) because of the extraordinary length of stage 2 (180 days). This stage begins in mid-autumn as the temperature falls below 5°C. The small flower buds therefore remain dormant for some months until fresh growth results in flowering by mid-spring.

LATE GROUP

Only one of the late varieties, Mackay pea, develops the most rapidly at Saskatoon and for the same general reason as the three early ones. In contrast the other two varieties develop the most *slowly* because key individual responses reverse the situation.

The late Mackay pea has an MG with a strong R5. However, the growing period at Saskatoon is shortened to nearly one-third that at Melbourne mainly by the universal responses to the higher temperature. These responses on their own would halve the growing period by shortening the three stages. The longer photoperiod at Saskatoon affects R5 and thus accounts for the even faster rate. Earlier FI lowers the VN from 32 to 17 and thus shortens the first stage a little further than through the action of R1 alone.

At Aberystwyth this variety, like the early one, dies because of low temperature in autumn, but probably while still vegetative because of the shortening photoperiod.

In Kharkov wheat the MG includes a slower tendency to flower than the early Australian varieties (Gabo and Insignia) and also the moderate R3, R5 and R6 as in other cereals. The growing period is doubled at Saskatoon compared with Melbourne (from 218 to 430 days). The developmental stages of Kharkov in the two types of season are essentially of the same length as for the Holdfast variety when sown in spring and in winter at Melbourne (Fig. 8.2). Reference to Table 8.1 shows that the second stage is over five times as long as at Melbourne while the other two stages differ much less. In spite of the long photoperiod at Saskatoon, the high summer temperature affects R3 and delays the first stage. The second stage therefore is likely to begin in late summer and is retarded by lowering photoperiod and then by the cold autumn and severe winter. In contrast at Melbourne, after the sowing in winter and FI in early spring, the reproductive apices develop fast in the lengthening photoperiod and rising temperature.

In Tallarook subterranean clover the MG includes a very strong R3 which prevents FI above a threshold temperature (see p. 113). Hence at Saskatoon, because of the high temperature after sowing in spring and despite the long photoperiod, the growing period is much longer than at Melbourne when the plants are sown in winter (only about 200 days).

The growing period at Aberystwyth is also longer than at Melbourne (308+ days). The warm temperature after sowing and then the long cold winter delays FI until late winter, and the cool spring also slows R1, thus lengthening the second stage.

4 Trends in Development with Latitude

How does latitude affect rate of development when the time of sowing is the same? This question is examined in two parts: the first considers plants sown in winter from middle to low latitude, and the second, those sown in the warmest part of the year from middle to high latitude.

WINTER SOWINGS FROM TEMPERATE TO TROPICAL ZONES

Plants sown at the same time in a range of places from 38° to 14°S in Australia grow in higher temperatures with decreasing latitude but the trend

Table 8.2

When sown in winter in various latitudes in Australia, maturity character in pea, wheat and oat interacts with the temperature and photoperiod† to affect the development rate in selected varieties

			Melbourne	Brisbane	Alice Springs	Ayr	Kununurra	Katherine
Latitude °S			38	27	24	20	15	14
Temperature °C			10	14	12	17	23	22
Photoperiod hr/dy			10·5	11·5	11·7	12·0	12·5	12·5
Maturity	Species	Variety	Measures of development					
			Days for development stages 1 and 2					
Early	pea	Collegian	91	64	67	46	52	41
Late	pea	Mackay	123	148	**	—	**	83+*
Early	oat	Alpha	140	100	85*	85	95*	90*
Late	oat	Algerian	155	—	125*	—	145*	125*
Early	wheat	Insignia	120	95	85	80*	85*	70*
Late	wheat	Odin	185	185*	**	**	**	**
			Mean number of vegetative nodes per main shoot‡					
Early	pea	Collegian	13·6	14·6	13·9	14·2	14·2	13·6
Late	pea	Mackay	34·5	47·9	**	—	**	37·2 few veg
Early	oat	Alpha	9·3	10·1	10·1	9·9	12·0	12·1
Late	oat	Algerian	11·8	—	12·9	—	—	20·0
Early	wheat	Insignia	8·6	9·6	9·5	9·8	10·4	10·7
Late	wheat	Odin	14·5	15·3	**	**	**	**

* some plants heading much later
** died before heading
† mean daily values of climatic factors, first month after sowing early winter, 5 June 1959
‡ coefficient of variation increased at lower latitude
veg vegetative plants

in photoperiod will depend on the season of sowing. Table 8.2 shows that the mean daily temperature in the first month after sowing in early winter is 12°C higher at Katherine in the Northern Territory than at Melbourne, but the photoperiod only increases by 2 hours.

Evidence from four species indicates that the trend in temperature is the more important one. Decreasing latitude is associated with faster or slower development depending on the dominance of the universal or the temperature-negative responses. Three of the species (pea, wheat, oat) are alike in the tendency for the early varieties to be faster and the late ones slower. These species will be considered first.

Early varieties of pea, wheat and oat

The figures for Collegian pea, Insignia wheat and Alpha oat express differences in MG. Because of its simple MG, pea reaches first flower at Katherine in less than half the time taken at Melbourne (from 91 to 41 days) with no change in VN. With wheat and oat the presence of slight temperature-negative responses slow development as shown by the smaller difference in the number of days from sowing to heading between the two places, and the higher VN at Katherine.

Late varieties of pea, wheat and oat

In spite of incomplete results, it can be seen that the late varieties develop more slowly than the early ones with decreasing latitude.

In Mackay pea R3 increases erratic development with higher temperature as shown by the larger variation of VN, and the strong R5 keeps the VN high at each place because of the moderate photoperiod.

In Algerian oat higher temperature activates the moderate R3 and almost compensates for the faster rate of leaf appearance resulting from universal R1. VN rises from 11·8 units at Melbourne to 20 at Katherine and the number of days from sowing to heading varies little between the cool and the hot growing seasons.

In Odin wheat the higher temperature at Brisbane activates the strong R4 and keeps the period from sowing to heading the same as at Melbourne, in contrast to the smaller effect of the weaker response in the early variety. In lower latitudes, at still higher temperature, the plants died before heading. This suggests that R4 may have killed the shoot apices after FI.

Early and late varieties of subterranean clover

In contrast to the previous species, subterranean clover includes varieties with a much stronger R3. Both early and late varieties therefore develop more slowly with decreasing latitude. Figure 8.4 shows that the vegetative stage lengthens as temperature increases from Melbourne 38°S to Kununurra 15°S. In the early variety the number of days from sowing to FI is increased fourfold or more (from 30 to 120+ days). The later variety is even more affected. At Brisbane 27°S compared with Melbourne the vegetative stage is about 20 days longer and the VN is more than doubled by the higher temperature (15° and 11°C respectively). The plants remain vegetative at Ayr 20°S and at Kununurra—that is, in temperatures higher than 16°.

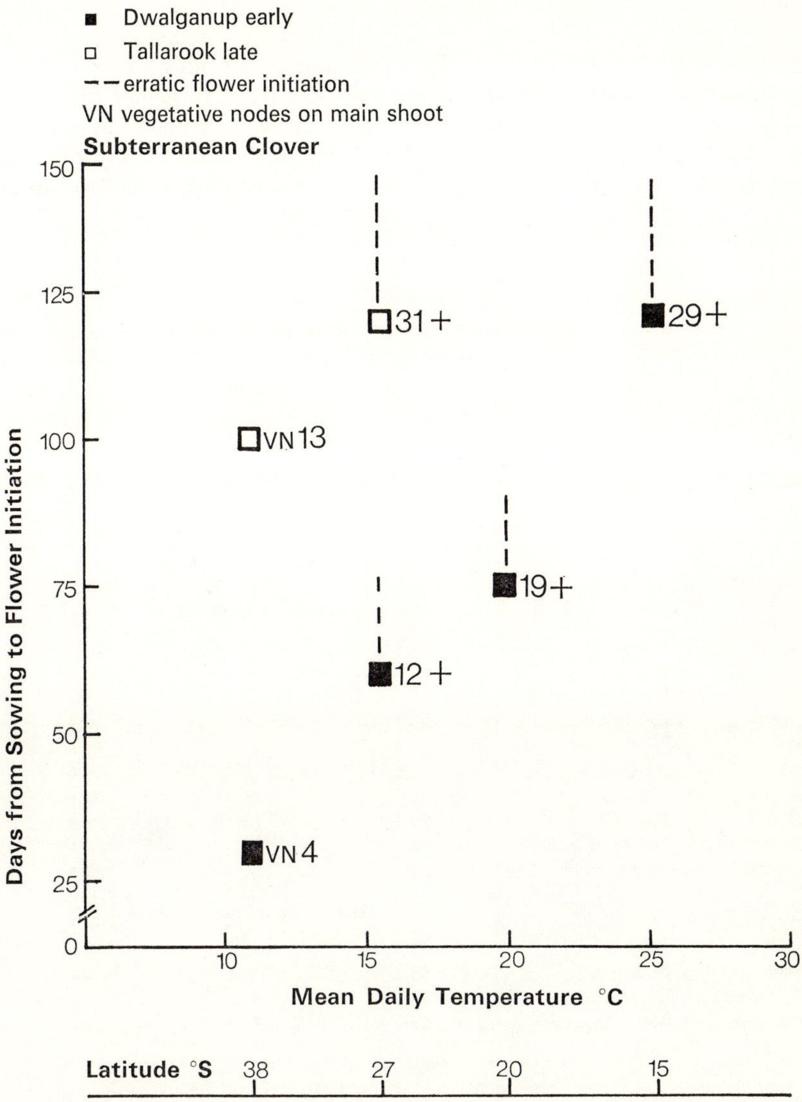

Fig. 8.4 In subterranean clover, both the early Dwalganup and the late Tallarook variety include the universal responses and the opposing temperature-negative R3. When sown in late autumn in a range of temperature in latitudes from 38° to 15°S in Australia, the higher temperature increasingly retards FI, as shown by the increased number of days from sowing and VN. This occurs in spite of increased rate of leaf appearance. In Tallarook the stronger negative response prevents FI entirely at mean temperatures above 15°.

SPRING–SUMMER SOWINGS FROM TEMPERATE TO ARCTIC ZONES

As latitude increases from the temperate to the arctic zone, plant growth is increasingly confined to the summer. This season is much cooler at very high latitudes than in lower ones, in spite of the high input of daily radiation during the few months of continuous light. The trend to cooler summers is stronger in coastal localities or small islands than inland on large land masses. Thus temperature may vary greatly according to location at the same latitude. In maritime environments, with increasing latitude, plants sown in late

Table 8.3
The effect of temperature on the development rate of early varieties of four species, in maritime climates ranging from temperate to arctic latitudes

		Site 3b Melbourne sown spring	Site 8 Macquarie I. sown summer	Site 10b Pt Barrow sown summer
Growing season†				
Temperature °C/dy		17·3 (100 days)	6·6 (100 days)	2·5 (52 days)
Photoperiod hr/dy for 30 days		15·3	18·8	24
Species	Variety	Measures of development		
		Days for development stage 1		
Pea	Alaska	7	15	20
Barley	Finlay	17	30	35+
Sub clover	Dwalganup	27	25	40+
Ryegrass	Merredin	50+	50+	45+
		Days for development stage 2		
Pea	Alaska	26	80	**
Barley	Finlay	29	65	**
Sub clover	Dwalganup	28	70	**
Ryegrass	Merredin	30+	50+?	**
		Days for development stage 3		
Pea	Alaska	31	*	—
Barley	Finlay	35	*	—
Sub clover	Dwalganup	45+	*	—
Ryegrass	Merredin	40+	*	—
		Number of vegetative nodes on shoot (VN)		
Pea	Alaska	8·6	8·6	9·0
Barley	Finlay	7·3	7·5	7·0
Sub clover	Dwalganup	8·3; 4·8‡	4·3; 2·1‡	4·0
Ryegrass	Merredin	13·0 veg	7·8	6·0

* no setting of seed because of frost?
** died because of low temperature?
† mean values of temperature and photoperiod after sowing spring or summer
‡ value on lateral shoot
veg vegetative plants over summer 30%
? estimated from the few plants that had reached final leaf when harvested at 95 days

spring or in summer not only begin growth in a longer photoperiod but also at a lower temperature.

Table 8.3 compares the growing seasons and the development of early varieties of four species at the three latitudes of 38°S, 54°S and 71°N. The mean daily temperature ranges from 17·3° to 2·5°, and the photoperiod from 15 to 24 hours—that is, the plants at the highest latitude (site 10b) compared with those at Melbourne (site 3b) grow at a much lower temperature (see Table 4.3) and at a photoperiod nearly 9 hours longer (Fig. 4.11).

The evidence shows that the trend of temperature is much more important than that of photoperiod in spite of the large difference in this factor compared with the small one with winter sowings from temperate to tropical zones. Increasing latitude is associated with slower development because low temperature retards the universal responses, but the more complex MG affects the trend slightly.

Early varieties of pea and barley

Because of their simple MG, Alaska pea and Finlay barley develop rapidly at Melbourne in the warm growing season, reaching ripe seed at 64 and 81 days respectively. In the cool growing season at Macquarie Island (site 8) development is two to three times as slow in the first two stages and fails after first flower because of frost damage. In the even colder season at Pt Barrow (site 10b), the first stage is a little longer than at Macquarie Island.

Early varieties of subterranean clover and annual ryegrass

In Dwalganup subterranean clover and the Merredin variety of ryegrass, individual responses retard the promotive effect of the warm temperature at Melbourne to some degree. These varieties reach ripe seed from 100 and 120 days onwards, mainly because their first stage is delayed by R3. Reference to the figures for the first stage at Macquarie Island compared with Melbourne shows equivalent lengths instead of those at Melbourne being only half as long. The fact that FI is delayed at Melbourne is also shown by higher VN than at Macquarie Island in both varieties.

Inspection of the second stage at the two places shows that it is lengthened at Macquarie Island as much in these varieties (that is, about three times) as in the previous pair. This indicates that R6, present in the Merredin variety, does not counteract the retarding effect of low temperature on R1 in spite of the longer photoperiod.

At Pt Barrow the varieties develop slowly to FI. Both the clover and the ryegrass include R5 in the MG but only the ryegrass provides evidence for a slight promotive effect of the continuous light. The first stage is about the same length as at Macquarie Island instead of being longer as with the clover, and the VN is several units lower.

General development in cool growing seasons

All varieties at Macquarie Island and at Pt Barrow failed to develop fully because the lower temperature slowed R1. This occurred in spite of any promotive influences that the longer photoperiod may have had on R5 and R6 in sensitive varieties.

At Macquarie Island the pea and the subterranean clover varieties had begun to flower and the barley to head at the harvest at 95 days, and the fourth variety (Merredin ryegrass) had reached its final leaf. However, further growth and development would have been prevented by lower temperatures in autumn. Other information shows that the ryegrass was the only variety of the four that was likely to have survived until the melting of the snow cover in the following summer.

At Pt Barrow all varieties had reached the early reproductive stage before the end of summer (52 days from sowing). Further development would have been prevented by rapid death in the ensuing, even colder weather.

5 Trends in Development with Altitude in Temperate Zone

In the tropics increased altitude lowers the temperature of the growing season as shown at the winter-sown plots in Hawaii (chapter 6). However, measurement of the effect of high altitude in the temperate zone is complicated by the length and severity of the winter. Active growth cannot begin until the snow melts in summer and it continues for only a few weeks before the snow covers the ground again. As altitude increases, the growing season begins later in spring and therefore may differ in both temperature and photoperiod from the lowest site.

Plants were grown in California at 38°N at the three altitude stations of the Department of Plant Biology of the Carnegie Institution of Washington (Table 8.4). The first two stations are alike in the warmth of their growing seasons in the three months after sowing while the highest station is cooler. The mean daily temperature is about 16°C at Stanford (altitude 30 m), slightly higher at Mather (1400 m) and much lower at Timberline (3050 m) as shown by 17° and 11° respectively. The mean photoperiod of the first month varies less: 14·4 hours at Stanford, 15·8 at Mather and 14·7 at Timberline (Fig. 4.9). Two other points concern the growing seasons at the three places. At Stanford the temperature of the first three months rises steeply from 12° to 18°, but at Mather it begins at 15° and then remains steadily at 17°. The growing season at Timberline is not only 6° cooler than at Mather but also much shorter.

Evidence from early varieties of four species shows that altitude affects development through temperature rather than photoperiod. The varieties develop more slowly at Timberline (site 4c) than at the other sites, mainly because of the slower universal R1, but other responses modify the rate slightly according to MG.

DEVELOPMENT AND MG

Each variety develops fairly fast to ripe seed at both Stanford and Mather but the three at Timberline only reach FI a little before the season ends with damaging frosts.

Table 8.4
The effect of temperature on the development rate of early varieties of four species, grown at different altitudes in California at 38°N

Growing season† Temperature °C Photoperiod hr/dy		Site 4a Stanford	Site 4b Mather	Site 4c Timberline
		15·7 (90 days) 14·4 (30 days)	17·1 (90 days) 15·8 (30 days)	11·0 (50 days) 14·7 (30 days)
Species	Variety	Measures of development		
		Days for development stage 1		
Pea	Alaska	13	8	*
Wheat	Gabo	36	24	40
Sub clover	Dwalganup	28	34+	45
Ryegrass	Merredin	30	40+	52
		Days for development stage 2		
Pea	Alaska	31	30	—
Wheat	Gabo	28	35	**
Sub clover	Dwalganup	22	30	**
Ryegrass	Merredin	22	30	**
		Days for development stage 3		
Pea	Alaska	33	32	—
Wheat	Gabo	37	41	—
Sub clover	Dwalganup	50+	64+	—
Ryegrass	Merredin	48+	50+	—
		Mean number of vegetative nodes on shoot (VN)		
Pea	Alaska	8·4§	8·7§	*
Wheat	Gabo	8·6	7·2	6·6
Sub clover	Dwalganup	8·9§ 3·8‡§	11·2§ 6·9‡§	5·5
Ryegrass	Merredin	13·5§	12·0§	6·0

* eaten by ground squirrels
** died from frost?
† mean daily values of temperature and photoperiod after sowing spring or summer
‡ value on lateral shoot
§ mean values of vegetative nodes not significantly different between sites
the varieties of subterranean clover and ryegrass have a large coefficient of variation at sites 4a and 4b

An early pea

Because of its simple MG, the early Alaska pea reaches ripe seed at Stanford and Mather at about the same time, at 77 and 70 days respectively (Table 8.4). The slightly faster rate at Mather reflects the warmer beginning to the growing season, which shortens the first stage a little (to 8 days compared with 13 at Mather).

An early wheat

The early Gabo variety has the same growing period at Stanford and Mather (about 100 days), because individual responses compensate for the general effect of temperature on leaf appearance and seed maturation. Stage 1 is shorter at Mather than at Stanford (24 days compared with 36) both because the higher temperature increases R1, and the longer photoperiod, through R5, lowers VN (from 8·6 to 7·2). However, stage 2 is longer (35 days compared with 28) because the continued warmer weather after FI activates R4. Stage 3 is also a little longer (41 days compared with 37) because, after heading, the lowering temperature in late summer at high elevation slows R2. Higher still, at Timberline, stage 1 is almost doubled in length compared with Mather (from 24 to 40 days) because of the retarding effect of the lower temperature on R1.

Early varieties of subterranean clover and annual ryegrass

Both the Dwalganup subterranean clover and the Merredin ryegrass include R3 in their MG and this is the main reason for their longer growing period at Mather than at Stanford. They illustrate the importance of temperature in the first days after germination to the expression of this response.

Because of higher temperature at Mather compared with Stanford during the first month (15° and 12°), FI was retarded as shown by the longer first stage and by the higher values for VN. In addition both varieties were more erratic in time of flowering and in VN.

At Timberline stage 1 in both varieties is longer than at Mather because of the low temperature. However, the difference in length is fairly small because the warmer temperature at Mather also retards development in these varieties through R3.

IX Categories of Genotype and Growing Season

The previous chapters have shown that both temperature and photoperiod may vary greatly within and between growing seasons. However, the growing period of the one plant variety may or may not be strongly affected. The important question is whether the concept of categories of growing period according to temperature of the growing season can be applied to varieties in all types of growing seasons, even artificial ones.

To answer this question it is necessary to review both the growing seasons from which developmental information has come and the interaction of MG with growing season. The possibility is then discussed of using four categories of growing period as a general basis for organizing the description of MG.

1 Diversity of Growing Seasons

As outlined in chapter 4, plants are able to reproduce themselves in many climates. The period available for growth varies greatly in the intensity of climatic factors and in the length of time over which they apply. A range of these growing seasons has been used to study rate of development in a few species. Before reviewing the key points of these studies, it will be helpful to summarize the growing seasons on the basis of the values of temperature and photoperiod for the first month of the plant's growing period. The real situation in a particular locality is usually more complex because of the changing values of the climatic factors and because of different lengths of growing season.

Three other points must be remembered. Firstly the natural growing season in most agricultural and pastoral areas is limited to the most favourable part of the year. Survival of perennial species from year to year often depends on varietal resistance to severe stresses at other times.

Secondly a crop or pasture plant in the field is usually sown a little later than the actual beginning of a limited growing season. Thus the length of time available for the plant may be shorter and also the climatic factors at sowing may have changed in value since the beginning of the natural growing

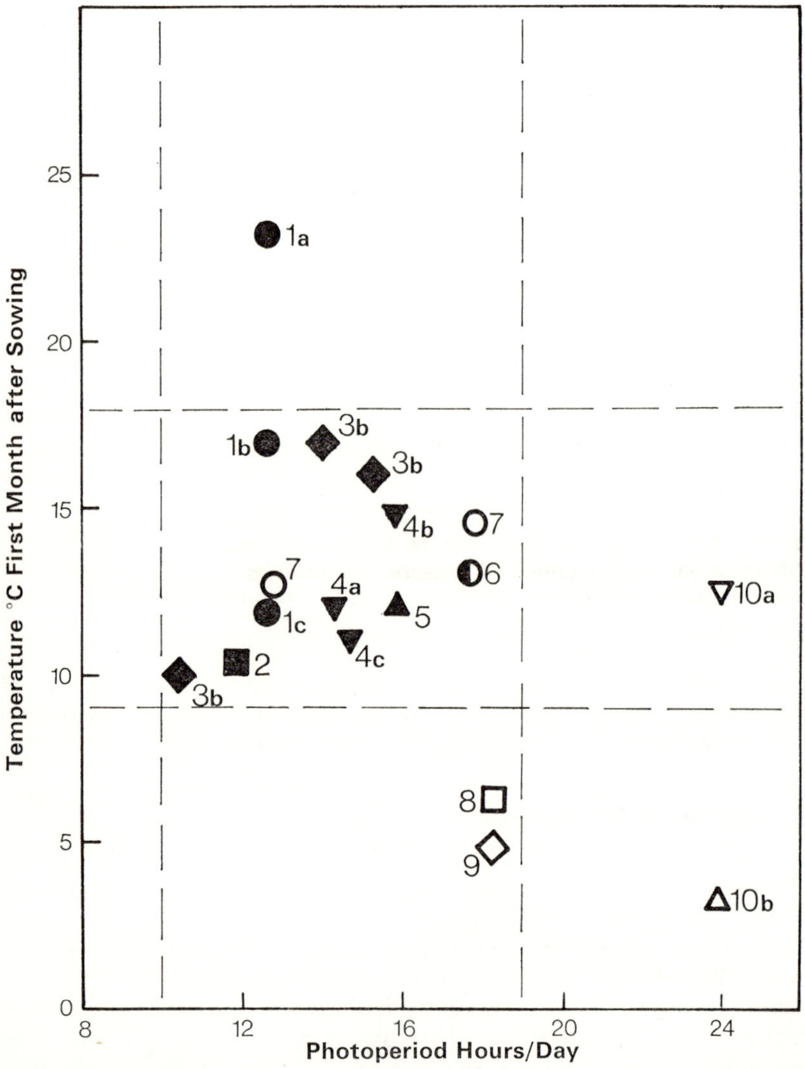

Fig. 9.1 Because the farmer tends to wait until both temperature and moisture become favourable for sowing, crops begin to grow within a fairly narrow frame of values for temperature and photoperiod. When arranged on the basis of these factors in the first month after sowing, most of the experimental sites cluster between 10° and 17°C and between a 10- and 18-hour photoperiod (see also Table 2.2 and Appendix 2).

season. Thirdly glasshouse horticulture may be carried on in 'artificial' combinations of temperature and photoperiod. These should be assessed in relation to the requirements of the species and varieties concerned and to the cost of providing these requirements.

GROWING SEASON AND CLIMATIC FACTORS IN FIRST MONTH AFTER SOWING

Reference to Figure 9.1 shows that the temperature of the month after sowing at most of the sites is confined to a fairly narrow band between 10° and 17°C, and the photoperiod for the same period lies between 10 and 18 hours. Only two lots of experimental sites lie outside the area. The site 1a at Hawaii (near sea level) represents the hot tropics with a photoperiod close to 12 hours throughout the year. The two sites with the longest photoperiod are typical of the northern boundary of arable farming (at College, Alsaka) and the cold margins of pastoral use (the coastal tundra at Pt Barrow, Alaska). The two sites close to 5°C are on the islands of Macquarie and South Georgia.

NATURAL GROWING SEASON AND AGRICULTURAL USE

Because of the plant's developmental responses (Table 5.2), it is interesting to note that an annual crop species is likely to be sown in many agricultural areas in the world when the temperature is close to or already above 10°. Growth may have become possible some weeks earlier because of rising temperature or the breaking of drought, but managerial reasons such as unfavourable soil structure or the need for weed control may delay sowing.

In contrast, perennial pasture species, once established, may begin to respond to climatic factors much earlier. For instance the early variety of red clover when sown in spring in Wales, develops very quickly in the long photoperiod and flowers 90 days later (mid-August). However, after overwintering, these plants would flower even earlier in the next summer (late June) because the lateral shoots, formed late in the first summer, are able to survive the winter in the vegetative stage. Thus they respond to lengthening photoperiod as soon as the snow melts in the early spring.

2 Interaction of MG with Growing Season

The expression of the one MG in different growing seasons depends on its responses to temperature and photoperiod rather than on the minimum VN which is more independent of the environment.

The universal responses of the variety provide a positive relationship of temperature with rate of development. The relationship may be reduced or intensified by the presence of individual responses to temperature or photoperiod, by the timing of exposure to these factors and by the intensity of the responses.

GROWING PERIOD AND DEVELOPMENTAL RESPONSES

The variety with only universal responses

Given that the seed emerges above ground safely and that moisture, nutrients, light intensity and temperature do not limit growth, the rate of de-

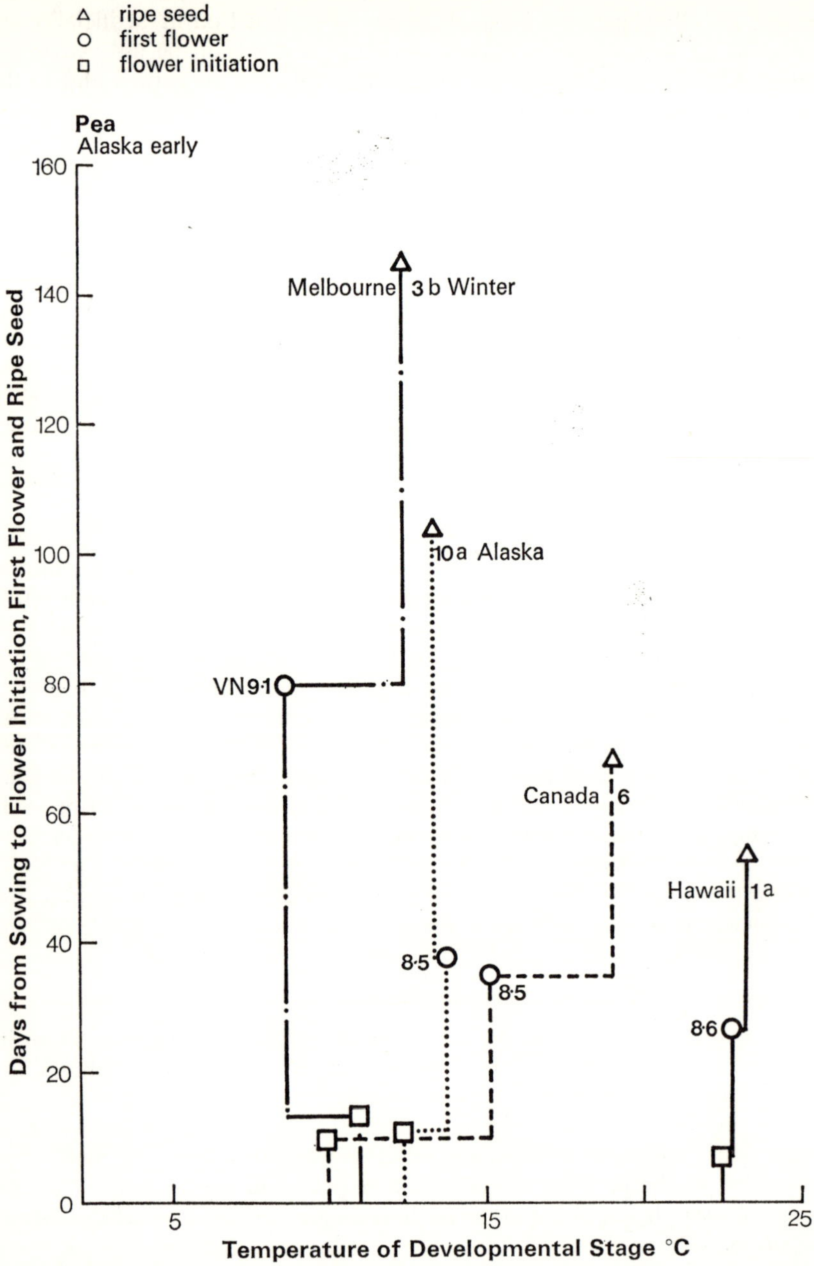

Fig. 9.2 The early Alaska pea includes only the universal responses to temperature in its MG. Although growing seasons may differ in both temperature and photoperiod, the growing period of this variety responds only to the temperature as shown by the trend to shorter length as the factor increases in value at the four sites. All three stages show the same trend. Each growing period is shown as a line made up of the lengths of the three stages, each placed vertically above its mean daily temperature.

velopment is positively correlated with temperature, irrespective of photoperiod, in all those varieties which have only or almost only the universal responses. In the early pea (Fig. 9.2) the length of each stage of development is inversely related to the mean temperature for that stage in each of four growing seasons. Two of these scarcely change in temperature: the hot season at site 1a in Hawaii and the cool part of the winter-spring season at site 3b at Melbourne. In the third growing season at site 10a in Alaska the mean temperature for each of the three stages also changes little, so that the line diagram for the variety remains at about 13°C although the temperature of stage 3 falls towards the end. In the fourth growing season at site 6 in Canada the temperature rises fast so that the line diagram for the three stages shifts to the right with each stage.

The figure shows that the growing period is more than halved as the temperature increases from about 10° to 22°, mainly because of the same trends in both stages 2 and 3. The length of stage 2 (FI to first flower) is about 70 days at 9°C at Melbourne, close to 30 days at 15° in both Alaska and Canada, but only 20 days at 23° in Hawaii. Stage 3 (first flower to ripe seed) takes about 70 days in the cool late summer in Alaska and the early spring at Melbourne, but only 30 days in the hot mid-summer of Canada and in the hot winter of Hawaii near sea level.

The variety with individual responses

In other varieties the presence of individual responses in the MG may affect the length of a stage directly or indirectly and thus change its general relationship with temperature.

Figure 9.3 gives an example of the way the general inverse trend of growing period with temperature may be masked by the action of the temperature-negative R3 in the early subterranean clover. Three of the growing seasons are the same as in the previous example (Fig. 9.2). The fourth one (site 3b at Melbourne, summer sown) replaces that at site 10a, Alaska. The growing season is also warm at first and then becomes cool; however, the temperature is higher to begin with and the photoperiod shorter.

In two of the growing seasons, the pattern of development of the clover is similar to that of the early pea. The growing period is shorter in Canada than after the winter sowing at Melbourne, mainly because the higher temperature reduces both stages 1 and 2.

The two other growing periods are, however, abnormally prolonged. In Hawaii at site 1a the high temperature, through R3, strongly delays development in stage 1, thus lengthening it three-fold compared with the sowing in Canada (from 27 to 90 days). The growing period in Hawaii is therefore almost double the length of that in the cooler season in Canada.

The other long growing period follows the summer sowing at Melbourne, partly because the high temperature after sowing activates R3 and partly because the long stage 1 brings the next two stages into the lowest part of the seasonal temperature curve.

Comparing the winter sowing with the summer one at Melbourne, the length of stage 1 *increases* from 27 to an erratic 40 days. In a variety without

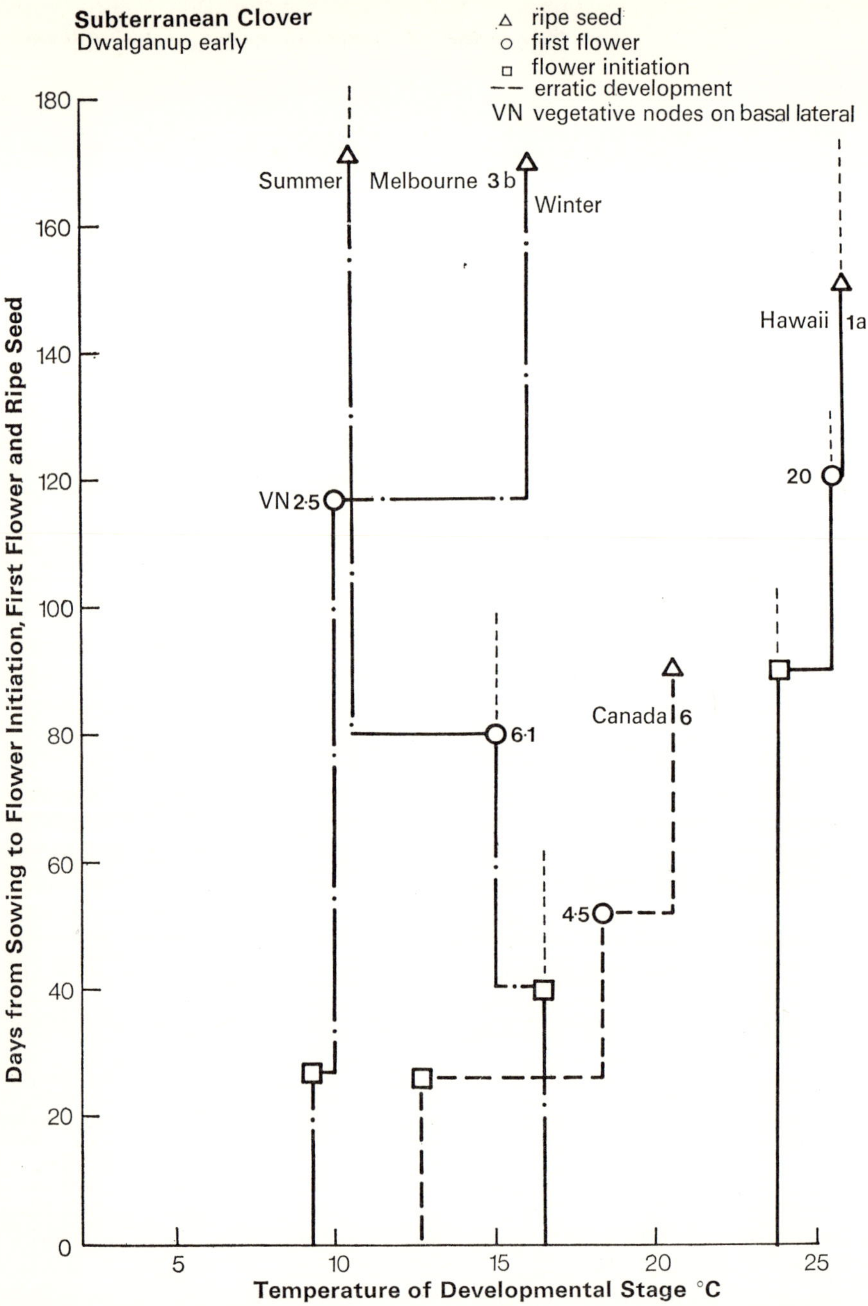

Fig. 9.3 The early Dwalganup subterranean clover includes not only the universal responses to temperature but also a strong negative response. Increased temperature shortens the growing period in seasons that are cool sufficiently long after sowing, but warmer later on, as shown for site 6 compared with site 3b, sown in winter. In contrast high temperature from sowing onwards at site 1a greatly lengthens the vegetative stage but shortens the other stages. Each growing period is shown as a line made up of the lengths of the three stages, each placed vertically above its mean daily temperature.

R3, however, the high temperature would have halved the length, thus permitting development to first flower and to ripe seed before winter.

DEVELOPMENT STAGES AND CRITICAL VALUES OF TEMPERATURE AND PHOTOPERIOD

In growing seasons with changing values of temperature and photoperiod, the time at which the plant is exposed may be of critical importance to the rate of development in varieties with responses additional to R1 and R2. It is also important to all varieties when subjected to extremes of temperature.

Stage 1

Importance of climatic factor at beginning of growth. Varieties with R3 and/or R5 are sensitive soon after germination, and a range of growing seasons provides many combinations of temperature and photoperiod. Favourable values bring rapid FI close to the first 20 days in Dwalganup subterranean clover (Table 8.1) and in Golden Rain oat (Fig. 7.1). Unfavourable values retard the change to the reproductive state as shown by a longer first stage and higher VN, or even may prevent the change.

In varieties with a strong response to one climatic factor, only a small difference in value may greatly affect the length of the relevant stage. Take the example of the early Cowgrass red clover with its strong R5 when spring-sown in Wales (site 7) compared with early summer in California at 1400 m (site 4b). The mean temperature for the first month after sowing (Appendix 2) is about 15° at both places but only in Wales is the photoperiod long enough (18 hr) to allow FI to take place within 30 days. In California the slightly shorter photoperiod of almost 16 hours prevents FI in the first month, after which lower values further prolong the vegetative stage until the following spring, about 340 days after sowing. Thus the first stage of development is extended to eleven times as long as in Wales because the photoperiod is slightly below the critical threshold for the particular R5 in this variety of red clover, when the seedlings emerge above ground.

Variation in length of stage 1. Because of differences in complexity of MG, some varieties vary more in the length of stage 1 than others when grown in a range of seasons. A large variation indicates a strong key response.

Reference to Figure 5.2 shows that the varieties of simple MG (Alaska pea and Early Wonder oat) have a stable VN while the more complex ones vary greatly according to the value of the main climatic factor to which they are sensitive, such as photoperiod for both varieties of red clover. The effect of such sensitivity on the length of stage 1 must be considered against the fact that even the variety of simple MG will take longer to reach FI at low temperature than at high because of the universal R1. The length is twice as long at 7° as at 17°C (Fig. 3.11) and because the sites provide a wider range of temperature, the ratio is extended closer to three-fold (Fig. 10.1, Alaska pea).

Table 9.1 gives the mean values for the length of stage 1 for each variety at the experimental sites and also the location of the growing seasons in which the rate of development to FI is the most extreme. In the early group

Table 9.1
MG, within and between species, interacts with growing season to affect development rate to FI, as shown by the mean length and standard error of stage 1 in varieties of nine species at the climate sites (number of growing seasons per variety is 13–15)

Species	Variety	Mean length of stage 1: days	standard error of mean*	Growing season with length of stage 1: least	longest
Early maturity					
Pea	Alaska	12	1·0	3b spring	3b winter
Oat	Early Wonder	19	1·6	3b spring	3b winter
Barley	Bankuti	22	2·1	10a spring	3b winter
Rye	Sth Australian	31	3·2	10a spring	3b summer
Wheat	Thatcher	39	4·2	10a spring	3b winter
Sub clover	Dwalganup	35	5·2	5, 7 spring	1a winter
Red clover	Cowgrass	104	21·3	10a, 7 spring	4b summer
Ryegrass	Merredin	59	9·6	10a, 4a spring	7 autumn
Ryegrass	Kangaroo Valley	61	9·5	10a, 4a spring	7 autumn
Late maturity					
Pea	Mackay	67	8·0	10a spring	3b winter
Oat	Algerian	62	7·4	10a spring	3b summer
Barley	Fimbul	92	6·6	10a spring	7 autumn
Rye	Varnerag	123	1·4	10a spring	7 spring
Wheat	Kharkov	151	1·5	1a winter	7 spring
Sub clover	Tallarook	197†	25·5	10a spring	lab winter 3b spring
Red clover	Montgomery	155	29·0	10a, 7 spring	4b summer
Ryegrass	Wimmera	68	10·2	10a, 4a spring	7 autumn
Ryegrass	Victorian	172†	31·3	10a spring	7 autumn lab winter

* The range of stage 1 for each variety varies with MG. The variety with the smallest standard error has a ratio between extremes in length of stage 1 of about 3, and the one with the largest error has a ratio of about 13.
† results omitted because all plants remained vegetative, for Tallarook subterranean clover at two sites and Victorian ryegrass at three

the oat and the pea varieties vary least because of their simple MG, and the red clover variety varies most because of its strong R5. In the late group, all varieties are complex in MG but those with moderate responses such as Kharkov wheat and Varnerag rye vary less than others with much stronger responses such as Victorian perennial ryegrass and Tallarook subterranean clover.

Stage 2

The length of the period between FI and flowering is controlled by the number and the rate of development of the vegetative phytomers which remain near the shoot apex above the top open leaf at the time of FI.

The rate of development is slowed not only by universal R1, as temperature falls from 23° to 7°, but also by other responses and by factors such as frost. Strong responses (R4, R6) to high temperature and to lowering photoperiod may slow both leaf appearance and differentiation of the reproductive shoot apex. Very low temperature may even kill the apex (Table 9.2).

Table 9.2
In subterranean clover, wheat and barley, MG strongly affects the length of development stage 2 through interaction with the temperature and photoperiod, after sowing in three growing seasons at two latitudes

Species and variety	Maturity	Length of period (days) Stage 1	Stage 2	Growing period	Temperature and photoperiod of stage 2 °C	hr/dy
		Growing season at site 7 (Wales), sown autumn				
Sub clover						
Dwalganup	early	28	180	263	6·3†	10·3
Tallarook	late	160	90	310	7·4	12·9
Wheat						
Gabo	early	45	*	—		
Kharkov	late	190	76	326	9·4	15·0
		Growing season at site 3b (Melbourne), sown summer				
Barley						
Cape	early	28	60–240*	270	9·7	13·4
Fimbul	late	116	150	310	11·3	12·1
		Growing season at site 3b (Melbourne), sown spring				
Wheat						
Gabo	early	27	38	100	15·9	15·3
Capelle	late	120*	240	430	12·1	12·6

* Gabo wheat died from frost damage after shoot elongation to 30 cm and LS8, early winter. In other varieties the main shoots died close to heading (Cape barley) or soon after FI (Capelle wheat) but the laterals grew vigorously.
† mean temperature winter months = 4·5°C

Thus, in spite of the favourable temperature for growth, the late wheat variety Capelle when sown in spring in Melbourne, and the early Cape barley when sown in late summer, only develop slowly. The second stage of development is greatly lengthened by the retarding action of R4 and R6 and forms a major portion of the growing period.

In cool growing seasons, such as the one that follows the sowing in autumn in Wales, the effect of the responses depends on species and variety (p. 118). The prostrate, early subterranean clover reaches FI about one month after sowing in autumn because the mean temperature (13·7°) only slightly activates R3. Then the temperatures soon fall below the level for active growth, and flowering is prevented until the spring.

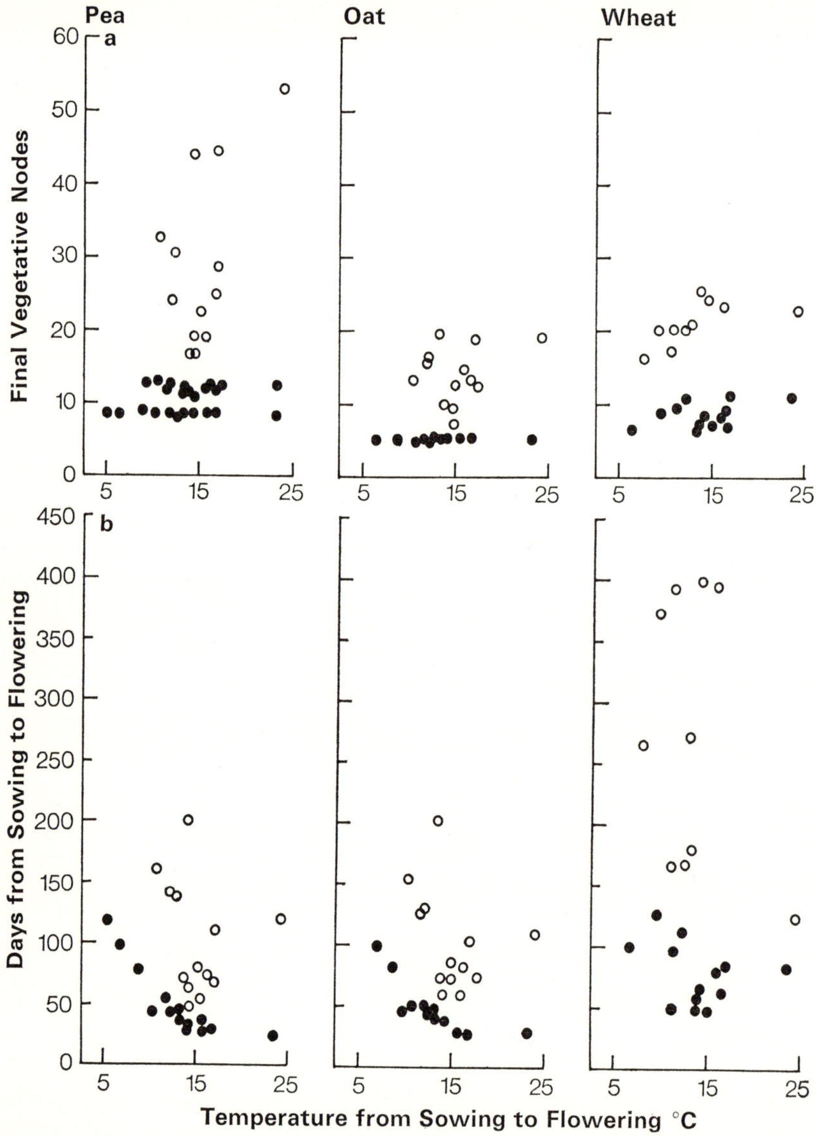

Fig. 9.4 When grown at the climate sites, the MG of a variety affected the relationship of temperature with the days from sowing to flowering (b) and (d) and with the final number of vegetative nodes on the shoot (a) and (c). The differences in pattern between early (solid circles) and late (open circles) varieties are shown for six species. The early varieties of pea and oat, with only the universal R1, take fewer days to flower with higher temperature, and the VN is stable. The late varieties

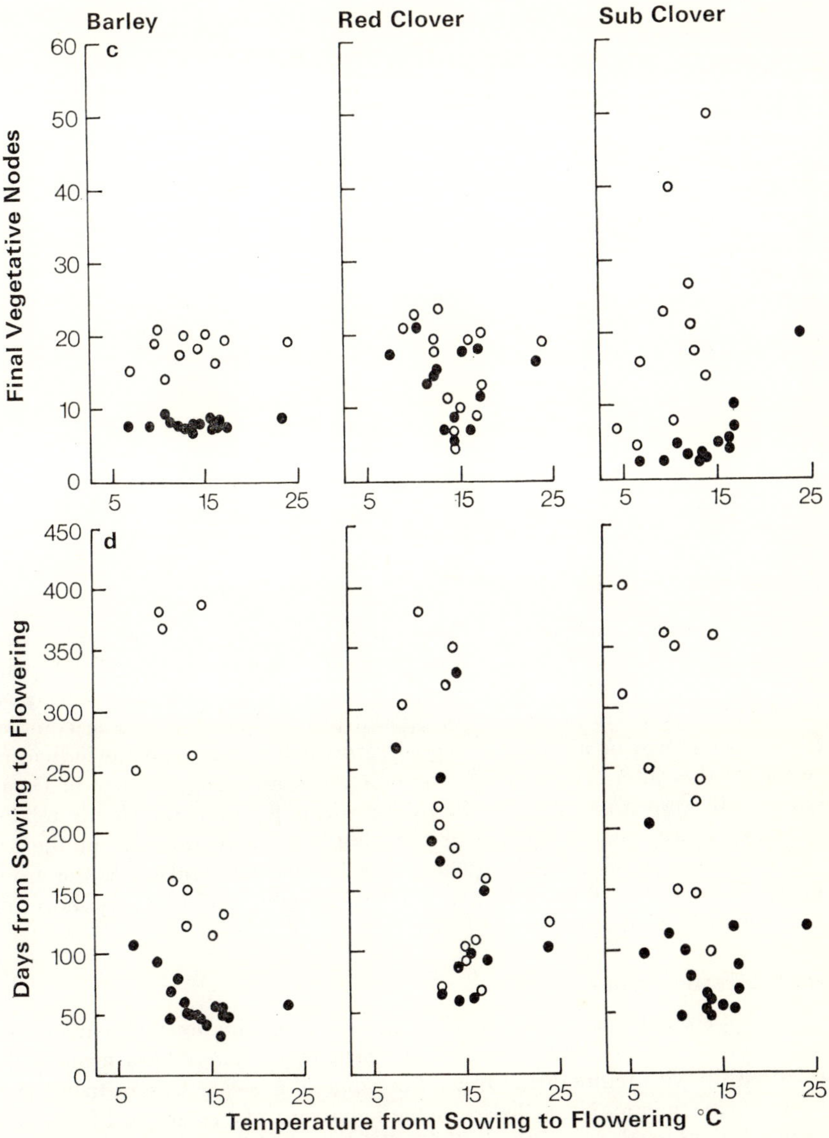

of these species and all other varieties are more complex. Early varieties: pea, Alaska, Collegian; oat, Early Wonder; wheat, Thatcher; barley, Bankuti; red clover, Cowgrass; subterranean clover, Dwalganup. Late varieties: pea, Mackay; oat, Algerian; wheat, Kharkov; barley, Fimbul; red clover, Montgomery; subterranean clover, Tallarook. Each circle represents the mean of ten plants at each site. The values for the vegetative nodes usually have a coefficient of variation of less than 2 per cent.

On the other hand an early cereal such as Gabo wheat also reaches FI soon after sowing but the young apex is killed by frost because of elongation above ground in late autumn. The late Capelle variety however survives because of its prolonged vegetative stage, during which the shoot apices remain at ground level and are thus protected under the snow.

Stage 3

The length of this stage is inversely related to temperature through R2, but may also be affected by the coincidence of stage 3 with extremes of temperature or with risk of other environmental hazards. Pollination may not even occur because anthers do not mature at too low a temperature or are killed by frost. Both situations affect cereals which head in winter after sowing in the autumn in Melbourne. Hot dry days at flowering may also kill the reproductive organs. This may affect early cereals sown in the spring in Melbourne. Red clover in southern Australia fails to set seed because pollinating insects are usually absent. Declining temperatures in late summer at high latitudes may so delay ripening that the immature seed is killed by the first autumn frosts as occurs with varieties of wheat and pea in Alaska (Table 5.5).

3 Categories of MG and Flexibility of Growing Period

The previous section indicates how the growing period of the one variety in a range of growing seasons may be affected by the interaction of MG with stage of development. In the simple MG the growing period is inversely related to temperature as shown for Alaska pea in Figure 9.2. The complex MG may change or annul the trend if growing seasons activate individual responses that alter the length of one or more stages. For instance in Dwalganup subterranean clover (Fig. 9.3) the growing period is not clearly related to temperature, mainly because of the plant's negative response at the hot site 1a in Hawaii. Here R3 lengthens stage 1 to 90 days and hence the growing period to over 150 days. In a simple MG, stage 1 would be very short and the growing period only about 70 days.

MG AND TEMPERATURE

A complex MG may interact with changing combinations of temperature and photoperiod to delay FI and/or the next stage more or less strongly. Its relationship with temperature in a wide range of growing seasons may therefore appear chaotic. Nevertheless, if the behaviour of the simple MG is kept in mind, the knowledge of the individual responses involved in each complex MG makes the erratic picture understandable.

MG and days from sowing to flowering

The importance of temperature may be illustrated first with varieties which are compared by means of the measures of VN and of days from sowing to flowering. Figure 9.4 shows that the relationship of the mean temperature of this period to early and late varieties in six species varies greatly with MG. The early varieties of simple MG (Alaska pea, Early Wonder oat) show an unchanging VN over the range of temperature and a curve of length of days to flowering which is inversely related to this factor. In the early variety of subterranean clover, the complex MG increases VN with increasing temperature and also the length from sowing to flowering. However, the latter pattern is blurred because other growing seasons lengthen the time to flowering for different reasons such as the low temperature after autumn sowing in Wales (Table 9.2).

In both early and late varieties of red clover, neither VN nor the days to flowering are associated with temperature because the key response (R5) in each MG is to photoperiod (see Fig. 5.2). In the late varieties of pea and of subterranean clover the increase of VN with temperature is partly due to R3 but the more scattered values for the days to flowering result from the complex expression of this measure in temperate growing seasons. This reason applies to both VN and days to flowering in the late varieties of oat, wheat and barley.

MG and growing period

The addition of the third developmental stage to the period from sowing to flowering strengthens the positive relationship of development rate to temperature. However, great variation in growing period may still remain because of the strength of delaying responses to the seasonal temperature and photoperiod.

Differences between kinds of MG in the importance of temperature to the growing period are shown in Figure 9.5. The examples include the varieties used in the previous section together with early and late ones in rye and in the ryegrasses.

As with the number of days from sowing to flowering, the growing period is only closely related to the mean temperature in the varieties of simple MG. In those of complex MG the growing period varies much more in length at the same mean temperature and tends to have a wider range overall. The reasons for the erratic values permit the classification of these varieties into categories as discussed in the following section.

FOUR CATEGORIES OF GROWING PERIOD

When the relationship of MG to growing season concerns both tropical and temperate zones, four categories are needed (Table 9.3). Two of the three useful in the tropics are of general importance (categories 1 and 3) and are based on positive and negative responses to temperature respectively. Category 2 is limited to growing seasons with moderate photoperiods (chapter 6). It provides for the varieties with weak R3 and R4; the growing period stays about the same length as temperature increases.

The fourth category includes those varieties in which the growing period is not related to temperature because of abnormal lengths of the first or second stages through a strong R4, R5 and R6. This category is needed for a variety with a response of little importance to growing period over a range of altitude in the tropics but of major interest in higher latitudes because of

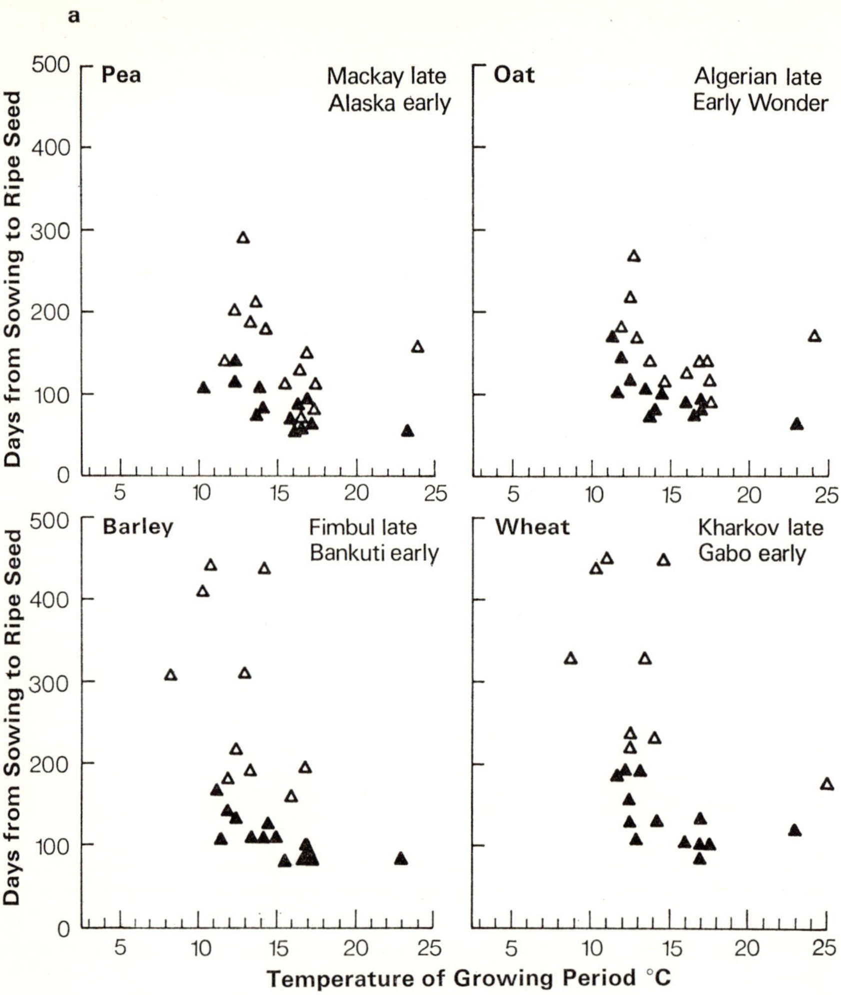

Fig. 9.5 When grown at diverse climate sites, the MG of a variety affected the relationship between the mean daily temperature and the days from sowing to ripe seed (growing period), as shown by the differences in pattern between early (solid triangle) and late (open triangle) varieties in each of eight species. The early pea and oat, with only the universal responses, show a clearer trend of shorter growing

variable temperature and photoperiod in the one growing season. Such a variety must be changed in category from that based only on evidence from the tropics.

For instance, at the three altitude sites in Hawaii the early red clover is put into category 1 because the period from sowing to flowering is reduced from 180 to 100 days with increasing temperature. However, other evidence shows that the variety develops slowly in Hawaii because the moderate photoperiod activates R5. In contrast, development is much faster at site 6 in Canada in the longer photoperiod (18 compared with 13 hr). Although

b

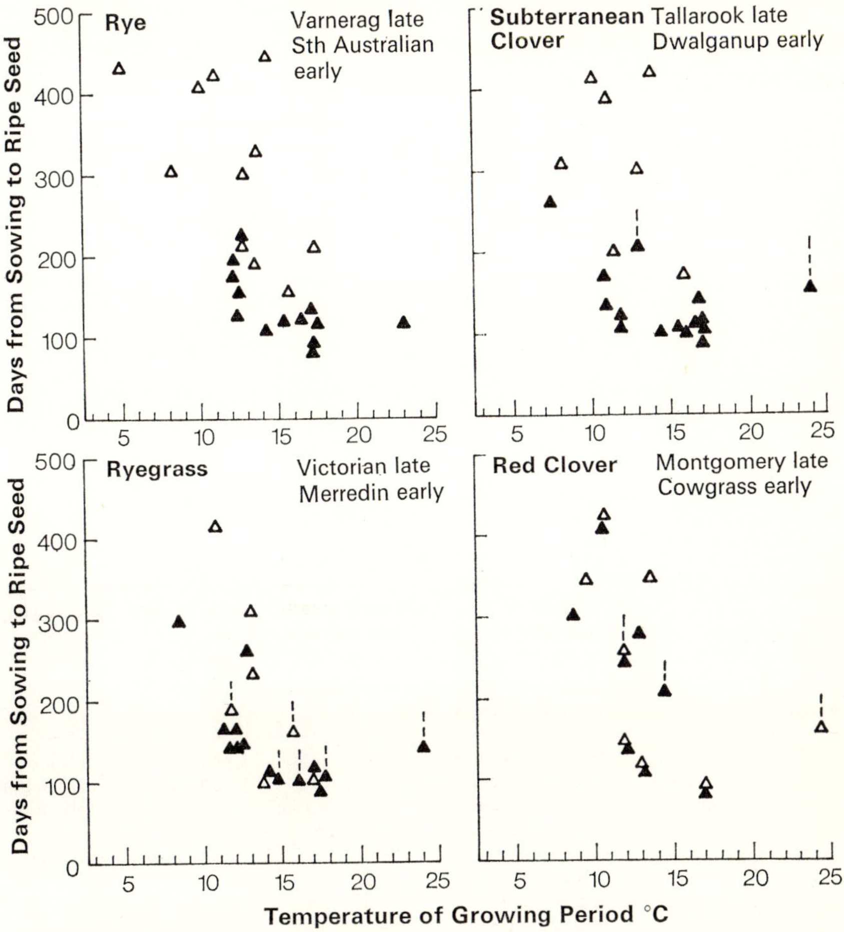

period with increased temperature than more complex varieties. Each triangle represents the mean value for ten plants at each site and time of sowing. A dashed line indicates erratic development. Red clover has few results because of lack of pollinating insects at some sites.

the middle altitude site in Hawaii has the same mean temperature (17°) the number of days from sowing to flowering at site 6 is reduced from 140 to 60 days and the VN from 13 to 7.

EXAMPLES OF CATEGORIES

The one variety in a wide range of growing seasons may or may not differ greatly in growing period. Examples of the extreme differences in growing period of varieties in three categories (1, 3 and 4) are shown in Figure 9.6. In each maturity group the varieties are placed in ascending order of

Table 9.3
When grown in differing seasons the MG of a variety may be classified into categories of growing period in relation to the temperature and photoperiod of the growing seasons and to plant responses (see also Table 6.4)

Category	Growing period related to seasons with increased: photoperiod	Growing period related to seasons with increased: temperature	Major plant responses other than R1 and R2	Species	Maturity
1	no effect or slight	shortened to half by rise 10–23°C VN stable or almost	none or only slight intensity R3R4, R5R6	pea oat barley	very early very early very early
2	(seasons with same photoperiod)	little changed by rise of temperature in moderate photoperiod more VN, longer stage 2, enough to balance effect of R1R2	moderate intensities R3R4	pea oat barley rye	mid, late early, late early late
3	small to moderate shortening	lengthened greatly to indefinitely with rise of temperature VN much higher	strong to very strong intensities R3, moderate R5	sub clover ryegrass perennial	early to late late
4	shortened by long photoperiod VN lower	may be shortened or else not related, because of abnormal length of stage 1 or 2 VN much higher and/or failure of first shoots in stage 2	moderate to very strong R5R6 and/or moderate R3R4	pea cereals red clover ryegrass annual	mid, late early to late early, late early, late

category with the lowest to the left side. Each variety is represented by two or three growing periods, which are arranged in order of increasing temperature of the growing season.

Early varieties

Category 1. This includes varieties of simple MG or with only slight responses otherwise. The simple MG has a narrow range of growing period (about 1:2), and the length of the period is inversely related to temperature only (Fig. 9.2). Figure 9.5 shows the typical curve of growing period with temperature in early pea and oat and a similar one in early barley which has a slightly more complex MG. In Figure 9.6, early pea and barley are compared. Both develop the most slowly when sown at Melbourne in winter but the pea is the most rapid in high temperature at site 1a in Hawaii while the barley is more rapid in Canada because of some sensitivity to photoperiod. Only the universal responses are involved, or almost entirely, as shown by the unchanged or almost stable VN in each variety (Table 9.4) and by the fact that increased temperature shortens the three stages of development.

Category 3. Early subterranean clover (Fig. 9.5) represents varieties with a strong R3 which tends to lengthen the growing period with increasing temperature. The growing period ranges from 90 to 250 days and the values are more scattered in relation to the mean temperature than in the varieties of category 1 (Alaska pea, Early Wonder oat). The growing periods in seasons with mean temperatures between 14° and 24° tend to lengthen with higher temperature while those between 7° and 14° tend to shorten because R3 is absent or much less active. However, other plants which develop in seasons within the cooler section of mean temperature are delayed more than expected. This is seen in Figure 9.6 with the early variety when sown in late summer at Melbourne (the left-hand column). Although the mean temperature is about 13°, the growing period is long (190+ days) because of seasonal variations in temperature. Development in stage 1 is delayed by the effect of high temperature on R3 and in the later stages by the effect of the low winter temperature on the universal responses (Fig. 9.3).

Similarly, the Dwalganup plants at some sites within the warmer section of mean temperature have a shorter growing period than expected. The second example of the variety in Figure 9.6 gives the growing period at the Canadian site. This is only 90 days in spite of a temperature of 17° because the mean temperature for the first 25 days after sowing is only 13°, thus preventing a strong expression of the temperature-negative R3. Stage 1 is therefore short and development fast in the other stages because of rising temperature. The third example in Figure 9.6 is that of the growing period at site 1a in Hawaii. Because of the steady high temperature, stage 1 is very delayed and erratic and is the major reason for the long growing period.

Category 4. Varieties in this category may vary greatly in growing period with little difference in mean temperature, because of the interaction between changing values of the key factor and the key response. Figure 9.5 shows that early red clover ranges in growing period from 90 to 410 days and also that in the lower part of the temperature range, very different growing periods have the same mean temperature. The MG in early wheat, rye and ryegrass includes more moderate responses and hence a somewhat smaller range in growing period.

In Figure 9.6 early Cowgrass red clover with a strong R5 is compared with early Cape barley with a weaker R5 and R6. In Canada (as shown by the right-hand column of each variety) the long photoperiod greatly hastens early development in both varieties in addition to the general reduction of all three stages by the universal responses to rising temperature. The growing period is thus reduced in each variety to one-fourth and one-fifth of the longest period respectively at the slowest site.

Table 9.4
When the one variety is grown in different seasons, the number of vegetative nodes on the shoot is related to the variety's category of response as in the examples in Fig. 9.6*

Species	Variety	Site	Sowing time	VN	Category
Early maturity					
Pea	Alaska	3b	winter	9·1†	1
		1a	winter	8·5†	
Barley	Bankuti	3b	winter	6·6	1
		6	spring	7·0	
Sub clover	Dwalganup	3b	summer	6·1§	3
		6	spring	4·5‡	
		1a	winter	19·7‡	
Barley	Cape	3b	summer	13·3	4
		3b	winter	9·8	
		6	spring	7·5	
Red clover	Cowgrass	4b	summer	21·0‡	4
		6	spring	6·8‡	
Late maturity					
Sub clover	Tallarook	3b	winter	8·0‡	3
		3b	spring	49·0‡	
		1a	winter	vegetative	
Barley	Fimbul	3b	winter	14·1	4
		3b	spring	18·0	
Pea	Mackay	3b	summer	45·0	4
		6	spring	18·3	
Red clover	Montgomery	4b	summer	23·0‡	4
		6	spring	8·8‡	

* the first sample of each maturity group corresponds with the left-hand one in Fig. 9.6
† the samples not significantly different
‡ the values are for basal laterals in red and subterranean clovers
§ this value is abnormally low because of an unusually cool late summer

The reasons for the longer growing periods differ according to variety and species. In Cape barley, the two examples of retarded development have the same mean temperature (about 12·5°) but differ in length. The plants sown in summer at Melbourne take longer (280 days), partly because R3 and R5 lengthen stage 1 in the high temperature and lowering photoperiod. Also R4 and R6 slow development in stage 2 as outlined for Insignia wheat (p. 115). In contrast the plants sown in winter take only 200 days, mainly because the lower temperature does not activate R4 in stage 2.

In the red clover variety, R5 is more sensitive to photoperiod than in Cape barley. The longest growing period (410 days) follows the sowing in summer at high altitude (site 4b) in California because the photoperiod is too low for FI until after the snow melts the following spring (see p. 135).

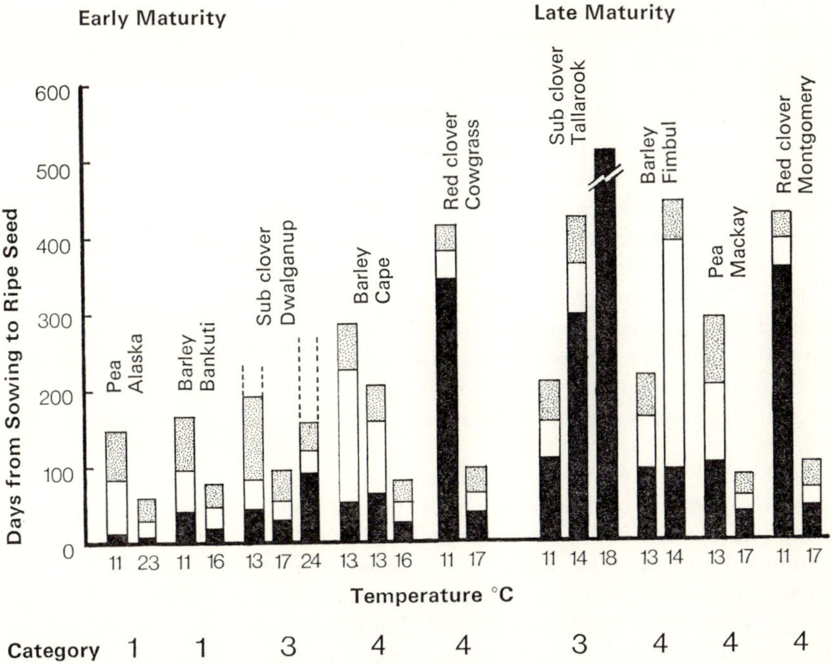

Fig. 9.6 On the basis of its development at diverse climate sites, a variety can be allotted to one of three categories of response to growing season, according to the relationships of the mean temperature (shown below each column) to the growing period and to the developmental stages. Within each maturity group, two or three examples of each variety illustrate the category. The varieties to the left of the early group are in category 1 because the universal responses shorten the growing period with increasing temperature. The varieties to the right of both maturity groups are in category 4 because responses to photoperiod dominate the length of the growing period. Each column shows for one variety at one site the three stages of development (vegetative, black; early reproductive, white; late reproductive, stipple). Erratic flower initiation is shown by dashed projections to the column.

Late varieties

Of the four varieties, one represents category 3, the others category 4. *Category 3.* In late subterranean clover, a very strong R3 prevents FI above a threshold close to 14°C. Figure 9.5 shows that the growing period varies between 200 and 420 days. Even more than in the early variety (p. 133) the temperature soon after germination is the key to the length of the growing period rather than the mean temperature of the growing period. The three longest periods follow sowings in spring and summer at the sites in Wales, Melbourne and California because the temperature is high enough to prevent FI until the following winter. Plants grown at steady high temperatures remain vegetative, as in Hawaii, and thus their values for growing period cannot appear on the diagram above the positions for mean temperatures of 17° and 24° (sites 1a and 1b respectively).

Figure 9.6 provides three examples to show the pattern of upward trend of growing period with temperature. The plants sown in winter at Melbourne develop the most rapidly, taking only 200 days because the low temperature after sowing (9° during stage 1) allows fairly fast FI without interference from R3, and the rising temperatures shorten the next two stages. In contrast the plants sown in spring take twice as long to reach ripe seed because R3 is activated by the higher temperature and the variety only becomes reproductive after the next winter. Sown in winter in Hawaii, the plants remain vegetative indefinitely at 17° and above.

Category 4. In Figure 9.5 it can be seen that late Fimbul barley, Mackay pea and Montgomery red clover have a wide range in growing period for little change in mean temperature. Fimbul barley varies from 160 to 440 days with the three longest periods after spring and summer sowings, as with the late subterranean clover. However, the reasons differ as shown in Figure 9.6, which compares plants sown in winter and in spring at Melbourne (site 3b).

In Fimbul barley in winter (left column) the plants take 220 days at a mean temperature of nearly 13°; both stages 2 and 3 are fairly short because rising photoperiod and temperature affect both R6 and R2. In spring (right column) the plants take almost twice as long but not because of the slight increase of mean temperature. Soon after sowing, high temperature activates R3, lengthening stage 1, after which both the high temperature and the lowering photoperiod strongly retard the young reproductive apices, through R4 and R6. Thus stage 2 is prolonged until healthy lateral shoots can head the following spring.

In Mackay pea (Fig. 9.5) the growing period varies from 83 to 290 days mainly because of the wide range of photoperiod and the strong R5. In Figure 9.6 the extreme cases are compared. The plants sown in summer at Melbourne (left column) are delayed in each stage, first because the lowering photoperiod affects R5 and then because the low winter temperatures slow the universal responses. Those sown in spring in Canada reach FI rapidly in the long photoperiod and continue to develop fast in the rising temperature.

In Montgomery red clover (Fig. 9.5) the range is from about 95 to 420 days. This is even wider than in the late pea but little different from the early Cowgrass red clover. In Figure 9.6 the comparison between the Montgomery variety sown in summer at high altitude in California and in spring in Canada is essentially the same as in the Cowgrass variety. R5 retards development in lowering photoperiod at Mather (site 4b) and hastens it in the long photoperiod at Saskatoon (site 6). In both cases, further development is rapid because of universal responses to the summer temperature.

X Application of Development Analysis

With the background of the preceding chapters, we can now look afresh at the amazing ability of plant species to flower and set seed in many climates.

How is the plant regulated to achieve flowering and ripe seed by a definite time? Can we use this knowledge to select varieties of desired species rapidly for our diverse agricultural areas?

The problem is examined from both the theoretical and the applied points of view.

1 Theoretical Aspects of Development Analysis

The present concept of MG helps one understand how plant development is controlled and points to further research.

BASIC PATTERNS OF GROWING PERIOD

Each MG is expressed in a phenotype that varies in rate of development according to the interaction of the genotype with the growing season. The growing period of the variety in different seasons forms a basic pattern with a minimum length and also a range in length. This pattern can be classified into one of the four categories of growing period according to its relationship with the mean temperature of the growing period (Table 9.3). For example Figure 10.1 shows the contrast between early and late varieties of pea. The early variety is placed in category 1 because it develops faster through each stage, with increased temperature. The shortest growing period is 60 days and the longest is only a little more than doubled. In contrast the late

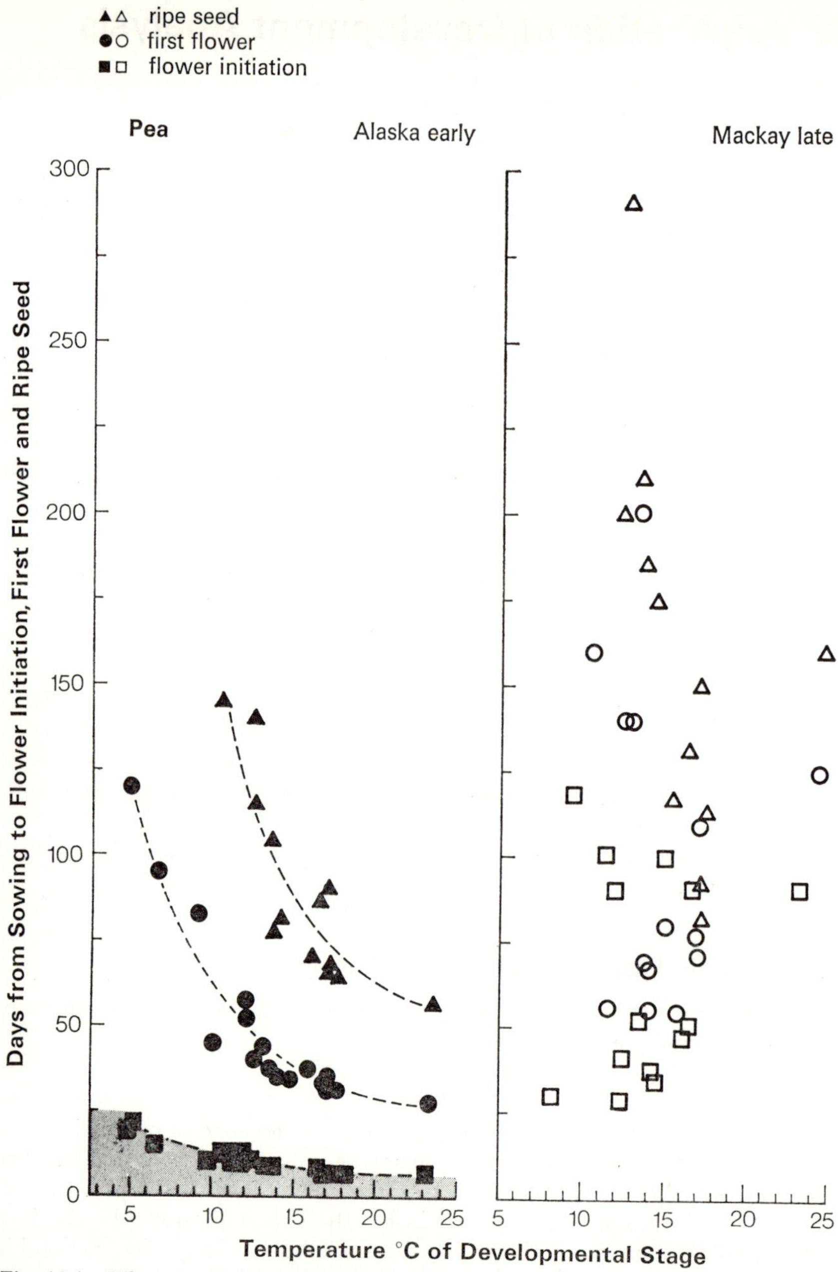

Fig. 10.1 When grown in a range of climate sites, the kinds of MG within a species differ in pattern of development as in the early and the late varieties of pea. In the Alaska variety the universal responses shorten the number of days to FI, first flower and ripe seed with increasing temperature, and the vegetative 'area' (grey) is well defined. In the Mackay variety the complex responses of the MG to both temperature and photoperiod give no apparent pattern of response.

variety belongs to category 4, because of the lack of pattern with mean temperature, the strong R5 in its MG, and the wider range in growing period, nearly 1:4.

RELATION TO NATURAL SELECTION IN CLIMATIC ZONES

In any one locality, well-adapted plant species are represented by populations suited to the growing season. The process of natural selection of appropriate MG is better understood when the genetic range in MG within each species can be translated into basic patterns of growing period.

For instance, consider the climates of the arctic and alpine tundras which have a very short, cool growing season in summer, followed by a very long cold period (Bliss 1956, Greene 1970). The environment favours the plant characters of perenniality, low growth, cold hardiness and early maturity. Almost all the native species are perennial and are represented by varieties in which the new shoot develops very rapidly in its first season to the early reproductive stage. The embryonic inflorescence may stay safely dormant for the next ten months under snow, progress a little further in the next growing season and become dormant again. The shoot may flower the next summer, setting seed in the more favourable seasons (Sørensen 1941, Hodgson 1966).

The MG in these varieties is likely to include a rapid tendency to flower (low minimum VN) and fast rates of leaf appearance and seed ripening (R7 and R8) to compensate a little for the retarding effect of low temperature. The temperature-negative R3 and R4 could also be present since they would be unexpressed in both types of cool summer. However, the strong photoperiod-positive R5 could be selected against in plants of the alpine tundras from low and middle latitudes. Active growth is likely to begin every year only from mid-summer as the photoperiod decreases (site 4c, California 38°S). Thus a population with a strong R5 could develop too slowly for the short growing season compared with one without the response.

In contrast the climate of the wheat belt in south-eastern Australia provides a growing season which is about six months over much of the area and includes a cool winter and a warm spring. The rainfall is uncertain; water stress is likely even in the growing season and drought is certain in the summer. The climate favours annual pasture species with an early MG.

The common annual species of ryegrass, *Lolium rigidum*, is represented by a heterozygous population (the Wimmera variety) that can regenerate by seed in spite of erratic seasons because of a small minimum VN and the presence of R3. This response does not retard development when delayed rains germinate the seed in winter but operates at the higher temperature of autumn when early rains permit a long growing season. Under these circumstances the plant grows vigorously in the vegetative stage, reaches flower initiation in winter, then flowers and seeds profusely in the spring. Thus the reserves of seed in the soil increase against short seasons in other years. Other kinds of MG would be inferior. A population without R3 would develop faster after an early break of season but might compete less strongly with other species in the pasture. Other populations could be too slow in development for sufficient seed yield because of a higher minimum VN and/or individual responses such as R3, R5 and R6 which are too strong.

Both the annual and the perennial ryegrasses in Australia provide examples of the natural selection of faster-developing populations from variable material in situations more difficult for slower varieties. On the drier edge of the wheat belt in Western Australia the growing season is often a little less than five months long and even more subject to moisture stress than the better country. The common annual ryegrass may still be an important part of the pasture if an earlier-maturing type is present. Established from seed of the commercial 'Wimmera' populations of the eastern states in the 1920s, the original, very small proportion of plants with a weaker R3 would have flowered earlier than the rest, set seed more readily and therefore soon predominated. This type of MG is the source of the early-flowering variety 'Merredin' (Reeves 1960).

In perennial ryegrass the Kangaroo Valley variety has an MG somewhat similar to that of the annual Merredin. The population is both early heading and perennial and is the local ecotype from pastures established last century in the south-east of New South Wales in an environment with a period of moisture stress in late spring. Later-flowering varieties such as Victorian fail to persist over summer, first because of soil dryness at heading and then because of competition with more active summer pasture plants such as paspalum (Aitken 1966b).

SURVEY OF GENETIC RESOURCES

Now that the MG can be described in terms of groups of units which provide basic patterns of growing period, a systematic survey is needed of the genetic resources in this character in the plant kingdom. The survey should include the temperate species of agricultural importance as one group and the tropical (or high temperature) ones as another. Within each group the range of MG within and between species could be compared in order to classify the variation into types of earliness and lateness.

The survey is likely to show that the units of the MG vary within every species. The temperate ones in the present study include up to six individual plant responses and six of the nine species each vary in minimum VN for early and late varieties. The species which are apparently without variation in the basic tendency to flower (red clover and the two ryegrasses) should be investigated further on this point.

The information for tropical species is still only partial. In maize, sorghum, soybean and cotton, maturity character and growing period are related directly to VN (Kuleshov 1931, 1933, Sieglinger 1936, Byth 1963, Ray 1966). The basic tendency to flower varies in maize (Francis 1969). Maize, sorghum and soybean respond positively to temperature (Holmes 1959, Quinby 1945, Steinberg 1936). Maize may also have a negative response (Duncan 1968) although this type of reaction is less likely to be of adaptive value to tropical species than to those native to temperate zones.

Photoperiod-negative responses within the species affect flower initiation more or less intensely in the 'short day' plants—maize, cotton and rice

(Francis 1969, Mauney 1963, Vergara 1967). Further, in both sorghum and soybean, these responses are more intense in later varieties (Quinby 1945, Steinberg 1936).

DEVELOPMENTAL GENETICS

The present knowledge of MG provides the basis for a study of the genetic control of rate of development in the flowering plant. The main questions are: how are the genetic units inherited and what are the biochemical and physiological processes of gene expression in the phenotype?

Interaction of MG and environment

This interaction holds the paradox that 'insensitive' varieties (that is, those with only the universal responses, category 1) develop more rapidly with higher temperature. The growing period therefore becomes shorter, as shown by Alaska pea in Figure 5.6. In contrast, 'sensitive' ones may vary greatly in rate of development because of individual responses, as shown in the same figure. The varieties in category 2 in the tropics are apparently unaffected by environment because their individual responses cancel the positive trend of the universal responses with temperature. Thus the growing period scarcely changes in warmer growing seasons compared with cool ones, as in Cape barley and Algerian oat. The varieties in category 3 develop more slowly at high than at lower temperature because of the dominance of the temperature-negative response, as in Dwalganup and Tallarook subterranean clovers.

The idea of categories of growing period is essential to the planning of research into the multi-unit nature of the maturity character. Varieties may be selected on their category and on their MG to be grown in the best environments for separation of the effect of particular genetic units. For instance, varieties in categories 3 and 4 include those with strong individual responses. Study of the effect of R5 on growing period can be made on varieties with and without the response if they are compared in long and shorter photoperiods at the same temperature. Varieties such as Golden Rain and Early Wonder oat (Fig. 5.5) differ dramatically in rate of development in moderate photoperiod and would be appropriate for such a study.

In later varieties the slower tendency to flower (late minimum VN) may be separated from the effect of individual responses by selection of the most suitable environment. This is illustrated from the times to flowering of early and late varieties in pea and oat (Fig. 10.2). The time from sowing to flowering in a number of growing seasons is related closely to the mean temperature of the period in the early variety but not in the late one.

In each species the basal segment represents the vegetative period of the early variety. The change in the width of the segment shows that flowering is reached faster with increased temperature. The width of the unshaded band represents the difference in minimum VN between the early and late varieties. It is expressed as the extra number of days taken by the late variety at each temperature when grown without activating individual responses.

The uppermost shaded area represents the remaining days taken for the late variety to flower because of the individual responses that retard FI in slow environments. In both species the figures for the curve of the early variety come from all growing seasons. However, to avoid complexity the information for the late one is restricted to extreme seasons. These include the three Hawaiian seasons which retard development and the Alaskan one which hastens the late varieties greatly (to 55 days from sowing for Mackay at site 10a and to 60 days for Algerian).

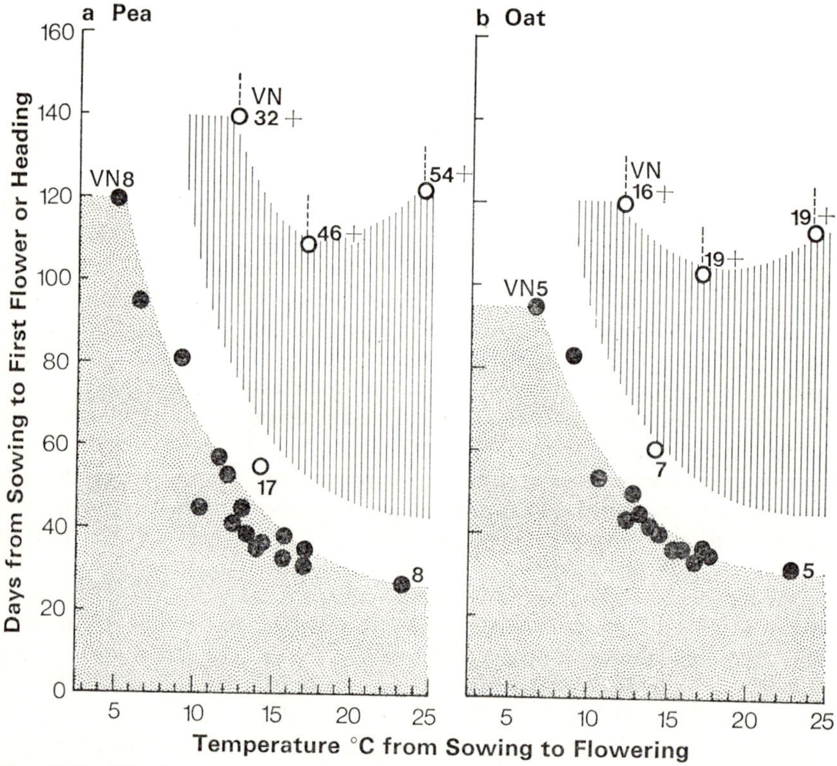

Fig. 10.2 The late-maturing characters in pea and in oat in any growing season result from delayed development beyond the time needed for the early variety, through a slower tendency to flower and individual responses to temperature and photoperiod. The contribution of the late character to the late phenotype is measured by comparing the relationship of temperature to the number of days from sowing to flowering in both varieties. In each species at all sites the early variety (stipple area) is affected by universal R1 only. The two areas above (white and barred) together indicate the extra number of days to flowering of the late variety in Hawaii at the three sites. The white band results from R1 and represents the time needed for the tendency of the late variety to flower more slowly. This is judged from its most rapid development in long photoperiod, as shown by Mackay (VN17) and Algerian (VN7) at site 10a in Alaska. The barred area results from the presence of responses in the late MG (Mackay: R1, 3, 5; Algerian: R1, 3-6). Each circle (first flower in pea and heading in oat) is the mean for ten plants at one site, and the adjoining figure indicates the number of vegetative nodes on the main shoot. Erratic flowering is indicated by a dashed line.

In the late pea the type of growing season not only affects the number of days from sowing to flowering but also the difference in length of this period between the late and the early variety and the proportion of this difference which is due to minimum VN and to individual responses. When grown at a temperature of 14°C with continuous light, as in the Alaskan summer 1970, the two varieties take 38 and 55 days respectively from sowing to first flower (Table 5.1). The difference (17 days) is small because, in Mackay, only the gene control for a later minimum VN is operating. At 17° in Hawaii the late variety is much slower than the early one (110 days compared with 35). The difference is large because, in addition to the slower tendency to flower, both R3 and R5 retard development in the moderate photoperiod at that temperature.

A comparison of the two oat varieties gives a similar picture for somewhat different reasons. The Early Wonder oat is like Alaska pea in its simple MG. However, the late Algerian oat has a less intense R3, R5 and R7 than the late pea but includes an additional one (R6) which affects the stage between FI and flowering.

Genetic analysis

The variation in MG raises questions that can only be answered by genetic analysis, and the inheritance of MG may be polygenic as shown by Clausen (1958a) in some races of *Potentilla glandulosa* and by Quinby (1945) in varieties of sorghum. The present study indicates that the MG includes several units, each of which could be controlled by a separate gene.

The first approach is to select the species that can best provide genetic differences in the desired unit of the MG. For instance, variation in the tendency to flower is part of the maturity character in pea, wheat and subterranean clover but is not obvious in red clover. One could widen the inquiry to include those species that typically have a prolonged juvenile period before the normal pattern of reproductive development begins. Are the gene systems that control the tendency to flower in herbaceous, short-lived species such as the cereals and pea, radically different from those in the woody perennials such as apple and plum?

The same approach applies to the individual responses. Subterranean clover can provide varieties with the temperature-negative R3 in a range of intensity. Cereals include variation within both R3 and R4 and wheat also has the modifier R8 in certain varieties. Similarly, appropriate species may be selected for the photoperiod-positive R5 and R6 (Table 5.2).

Given the knowledge of suitable species and varieties, the question arises as to the method for the genetic analysis. Should one concentrate on varieties that represent the extreme range of the maturity character in a species, such as Early Wonder and Algerian or Peniarth oat? Or should the varieties in a crossing programme represent a graded series in maturity because of increasing 'doses' of the one unit, such as that of R5 in the pea species? The success of the inquiry will rest on the choice both of varieties and of the environment so that the plants produce the maximum evidence for analysis.

Gene expression

Other questions concern the processes of gene expression and the best environment for their analysis. The genes controlling the expression of the maturity character may be latent in some environments but activated in others (Clausen 1958b). Thus in *Sorghum bicolor* (Miller 1968) eight varieties all began to flower in 50 days when grown in short photoperiod in Puerto Rico but revealed a range of flowering from 61 to 90 days in Texas because of varietal differences in photoperiod-negative responses to longer photoperiod.

The present studies show that the early and late red clovers flower together at long photoperiod because of their strong R5 (Table 7.1, Fig. 9.6) and they do not otherwise differ in tendency to flower. In subterranean clover the late variety has a strong R5 which only becomes activated as the mean temperature rises above about 12°C, but earlier varieties have weaker responses with a higher threshold.

We know that plants with an individual response in the MG are very sensitive to the value of the important climatic factor at the beginning of the relevant stage. The length of that stage may thus be determined within the first few days. The biochemical and physiological basis for such sensitivity could be followed in contrasting environments from the time of sowing of varieties with different intensities of response. An example would be the neutral Early Wonder oat compared with the Golden Rain variety with its strong R5 and R6 which retard development in moderate photoperiod (Fig. 5.5).

In what way do responses inhibit progress towards FI or from FI towards flowering? The first pair (R3, R5) affect the rate of development, independently of the universal response controlling rate of leaf appearance, but the second pair (R4, R6) interact with this rate. For instance, varieties such as Cape barley sown in summer at Melbourne soon reach FI in early autumn (Fig. 9.6, Table 9.4). However, they then become increasingly slower in rate of leaf appearance as well as in differentiation of the young inflorescence in the lowering photoperiod of autumn, in spite of continuing warm temperature for some time.

What is the basis for differences between species in the rate of leaf appearance (R7)? This response contributes to the slower rate of development up to flowering of early Bankuti barley compared with early pea, when sown at Melbourne in winter (Fig. 9.6). Contrasting varieties such as these could be studied but only in the vegetative stage in order to avoid conflicting responses that could be present in the later stages.

Similarly the differences between varieties of a species in the rate of seed ripening (R8) should be investigated only in those which are alike otherwise in rate of development, such at Kogat and Gabo wheat (Table 5.5).

2 Practical Aspects of Development Analysis

The basic problem of plant production in a given area is the fitting of desirable species to the environment to gain the optimum economic yield.

This means that varieties of the species must be selected that develop at the best rate for the growing season, according to the purpose for which the crop is intended. A plant may be valued for its seed as food, for its leaf as pasture or fodder crop or for its wood as timber. The growing seasons may be only three months long as in high latitudes or may continue throughout the year as in the most favourable climates. In the latter case the challenge may be to increase food production per hectare through the year, by a series of assorted crops, as is possible in the humid tropics (Bradfield 1969).

The selection of the best MG for a given growing season or portion thereof is best discussed from the aspect of the most common aim, that of seed production. However, the same principles also apply to crops harvested earlier, including those grown for leaf yield in the vegetative stage.

FORECASTING PLANT DEVELOPMENT

Knowledge of the MG of a variety makes the forecasting of rate of development possible in different growing seasons. The presence of the universal responses gives a positive relationship of development with temperature while other responses may strengthen or weaken the trend.

Temperature and category

The expression of the MG depends on its interaction with the temperature and photoperiod of the growing season. When the same variety is considered

Table 10.1
The relationship of growing period of three early varieties to global radiation and temperature in thirteen growing seasons

Measures	Species and variety		
	Pea Alaska	Oat Early Wonder	Barley Bankuti
Growing period			
mean days	89	97	107
standard error	8·4	9·5	8·4
Correlations (r)* *of growing period with mean daily values of:*			
global radiation (cal/cm²/dy)	−0·40	−0·27	−0·60
temperature			
1. mean	−0·82	−0·75	−0·74
2. maximum	−0·84	−0·80	−0·79
3. minimum	−0·60	−0·57	−0·51
4. effective maximum†	−0·87	−0·79	−0·79
5. effective minimum	−0·73	−0·66	−0·64
6. (mean + min) ÷ 2	−0·71	−0·66	−0·63
7. (mean − 5°C) × mean daily photoperiod‡	−0·90	−0·85	−0·85

* statistical significance: (P = 0·01) from r = −0·71; (P = 0·001) from r = −0·82
† Kimball 1959
‡ Nuttonson 1948

in growing seasons differing in both temperature and photoperiod, the growing period of the varieties in category 1 will be closely and negatively correlated with temperature (Fig. 10.1, Table 9.3). Those in the other categories will usually be less so. On the other hand, growing seasons that differ only in temperature may mask the full expression of the MG, as at the altitude sites in Hawaii. In red clover for instance, the effect of the photoperiod-positive response for each variety is similar at the three altitude sites but varies greatly in other growing seasons.

Table 10.1 shows the relationship of the growing period of three varieties in category 1 with different measures of temperature and with global radiation. Both mean and maximum temperatures (measures 1 and 2) and also mean temperature multiplied by photoperiod (measure 7) are significantly and inversely related to the growing period of these varieties in a range of localities. The association of measure 7 with growing period cannot be due to the stimulus of development by photoperiod because the varieties are free or almost free of photoperiodic responses. Instead the higher correlation occurs because measure 7 is related a little more closely to the effect of the daily total sum of temperature on the universal responses than the other

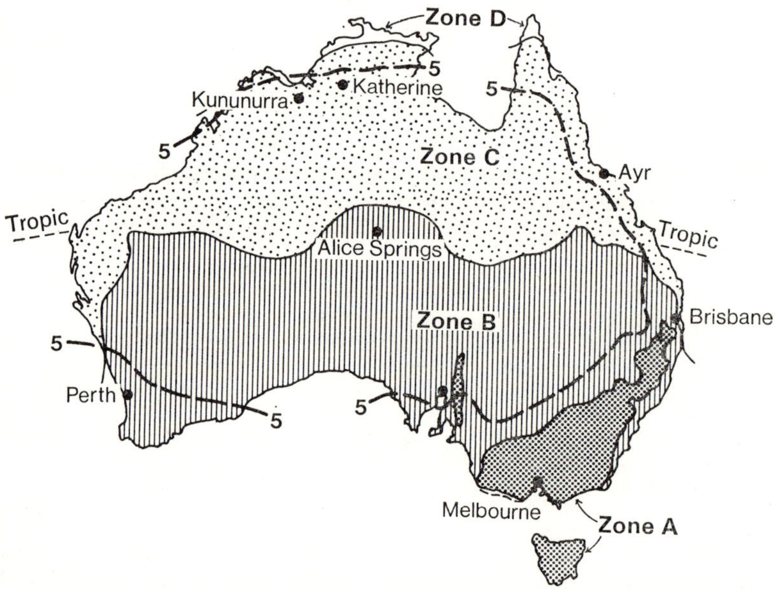

Fig. 10.3 The importance of the temperature-negative response to MG in subterranean clover justifies a map to show the probable relationship between the maturity character and the mean monthly isotherms for early winter (June) when the species is sown in winter in Australia, with irrigation where needed. The line marked 5 indicates the inner boundary of the five consecutive months or more of effective rainfall. Zone A (< 10°C), all varieties should flower normally; Zone B (10–15°C), all varieties should flower at higher nodes and reach FI more slowly; Zone C (16–24°C), early varieties should begin to flower at still higher nodes and later ones should remain vegetative; Zone D (> 24°C), even the early varieties should remain vegetative.

measures. The close association is likely to apply only when the universal responses dominate the rate of development.

Varieties in category 3 illustrate a different relationship of temperature with growing period. The growing period may be lengthened with increased temperature because of the inhibiting effect of the strong R3. This reaction is used to forecast the flowering behaviour of MG in subterranean clover in Australia, given adequate soil water. Figure 10.3 is based on the fact that the threshold temperature for the operation of R3 is lower for the later varieties than for the early ones. The plants are considered as sown in winter when the inhibiting effect of temperature on the response would be least during the year. In zone A in winter the temperature is below the level for activation. Both early and late varieties lengthen their first stage as the temperature rises between zones A and C, just as they do with decreased altitude in Hawaii. The later varieties remain vegetative in zone C, while the early ones are much delayed in flowering, thus preventing or reducing self-regeneration.

Flexibility of phenotype

A variety will vary in the length of its growing period according to the complexities of the interaction between the MG and the growing season. With sufficient knowledge a variety can be grouped into one of the four categories and classified on its minimum growing period and its degree of flexibility according to growing season. The latter characteristic may be illustrated by comparing the varieties in category 4 which vary greatly in growing period (p. 148) and those in category 2 which are scarcely affected in length by temperature in the tropics (p. 93).

Early measures of development rate

The relationship between the structure of the plant and the stages of development may be used to forecast its possible success in a given environment. Zaitzev (1923), who was selecting cotton varieties for the short growing season in Turkestan, found that the period from sowing to boll-dehiscence was highly correlated with the time from sowing to first flower. The evidence from the climatic studies shows that earlier measures of development may be used to assess the growing period of unknown varieties in a favourable environment. In Hawaii (Table 10.2) the growing period of varieties of cereals and pea are not only closely related to the days from sowing to flowering (stages 1 and 2) but also to the days from sowing to FI (stage 1), to the VN and to leaf stage at FI. When a variety has few plants, the latter measure is also important as a non-destructive alternative to dissection for FI (Aitken 1970).

An important aspect of plant structure is the relationship between the leaf stage of a species and the days from sowing which may be used to assess the rate of development of phytomers (Fig. 3.3). The rate of leaf appearance, as shown by the angle of slope for leaf stage, can be charted at different temperatures as in Figure 6.2. When the growing period must be limited to a certain length at a particular site, the leaf slope of a new variety can

Table 10.2
The relationship of growing period to other measures of maturity in pea and cereals at three temperatures in the Hawaiian Islands, 1963

Growing season	Site 1c National Park cool		Site 1b Kula warm		Site 1a Waimanalo hot	
Maturity measures and correlations (r)*	Pea	Cereals	Pea	Cereals	Pea	Cereals
Growing period						
mean days	145·0	178·0	113·0	121·0	76·0†	116·0
standard error	2·3	13·2	9·7	12·1	8·7	10·6
correlations with						
development stages 1 + 2	0·92	0·88	0·98	0·96	0·95	0·96
development stage 1	0·99	0·86	0·98	0·96	0·94	0·90
Development stages 1 + 2						
mean days	94·0	101·0	67·0	74·0	69·0	79·0
standard error	15·0	11·9	12·7	11·8	17·0	8·5
correlations with						
development stage 1	0·99	0·97	0·99	0·98	0·99	0·88
vegetative nodes	0·97	0·94	0·99	0·98	0·98	0·97
leaf stage at FI	0·99	0·96	0·99	0·92	0·98	0·93
Development stage 1						
mean days	46·0	55·0	42·0	35·0	44·0	38·0
standard error	12·6	11·6	14·4	10·9	14·1	9·9
correlations with						
vegetative nodes	0·95	0·95	0·99	0·97	0·98	0·95
leaf stage at FI	0·99	0·97	0·98	0·92	0·99	0·97

* the means are values ± standard errors of groups of varieties (pea 4–6, cereals 10–12) statistical significance: (P = 0·01) from r = 0·96 (pea), = 0·74 (cereals); (P = 0·001) from r = 0·99 (pea), = 0·85 (cereals)
† the two late varieties failed to set seed

be observed from the start of growth and the time projected to flowering according to the temperature of the growing season. Measurement of the leaf slope from sowing onwards as well as of the leaf stage at FI will help to show whether the variety is too slow in development.

Two factors must be kept in mind in this type of assessment. The first is that both rate of leaf appearance and rate of reproductive development after FI are positively related to temperature through universal R1. The second factor is the possibility, according to MG, that special responses may interact with both rates. R7 modifies the rate of leaf appearance according to species and thus affects rate of development up to flowering. R4 and R6 are temperature-negative and photoperiod-positive respectively and act during the stage between FI and flowering. These responses may retard or intensify the positive, universal response to temperature during this stage. R8 may partly counteract the retarding effect of low temperature on the ripening of seed.

AGRONOMIC SELECTION OF PLANTS

The understanding of the MG is important for plant introduction, breeding and production for seed.

Plant introduction

The problem of fitting variety to growing season is a difficult one. At first, one may have to select kinds of MG suitable for the local environment, knowing only their expression in a different one (Yarnell 1942). Skill is also needed to find the appropriate MG for other growing seasons, within or between countries.

The agronomist may be concerned with varieties from the point of view of equivalent growing periods as with irrigated winter crops in tropical northern Australia (Beech 1967). Or a method is needed by which varieties of different MG can be separated reliably at flowering, as with the early and later varieties of red clover when grown for commercial seed (Aitken 1964). The question may also arise as to how best to manage the variety in its early stages of development, as in cereals and pasture species when needed for both grazing and seed (Aitken 1961b, 1962).

In each case the task is assisted by analysis of the available growing season and if possible also of the growing period of adapted varieties of the desired species. Then estimates may be made of suitable MG and of likely sources of varieties.

Essential information. The growing season must first be assessed for its length and values of temperature and photoperiod. When considering the relevant growing season for an annual crop, one needs both the length from the normal time of sowing to harvest, and the mean daily values of temperature and photoperiod at intervals.

It should be remembered that temperature is the major factor affecting the full growing period through the universal responses and perhaps also through individual responses. On the other hand, photoperiod is much more limited. It affects only the first one or two stages in varieties with R5 and R6 and is a key factor in rate of development only in varieties with strong responses and in certain types of growing season. However, both climatic factors in the first few weeks after sowing may be crucial to the rate of later development, depending on the presence and intensity of individual responses.

The growing period of locally adapted varieties should also be analysed for rate of development, as measured by the number of days from the time of sowing to FI, to first flower and to ripe seed.

Selection of appropriate kinds of MG. Armed with the knowledge of the desired growing season, districts with similar growing seasons (homoclimes) may be located in which the crop is already cultivated. Introduced varieties from those places are therefore likely to have the appropriate rate of development. However, varieties from other types of growing season cannot be expected to be suitable locally. Instead they must be considered for their potentially useful kinds of MG. These may be deduced from the knowledge of their phenotypes in various growing seasons and introduced from the most suitable regions.

Geographical clues. The above deductions may be hindered by confusing terms for varietal development such as the 'vegetative period'. This may mean the time from sprouting to ripe seed (Kuznetzova 1928), from sowing to ripe seed (Vavilov 1923, Azzi 1956) or from sowing to heading or first flower (Nuttonson 1948). Floral initiation may mean the first obvious flower bud, as in cotton (Moraghan 1968) rather than the first reproductive sign of the 'double ridge' at the shoot apex. The term 'stage of development' may mean the appearance of each new node (or mature phytomer) on the shoot, as in the garden pea (Higgins 1952).

Hence dates of sowing, flowering and ripe seed are needed rather than less precise information. If the temperature during development is also known, the length of the third stage in the new environment can be estimated. In addition, one may be able to identify the type of individual response affecting the length of the first two stages. These points are illustrated in Figure 10.4 from combined information on three wheats of similar MG. The Gabo and Thatcher varieties are from the author's climatic studies and the Prelude variety is from Vavilov's geographical tests in Russia (Kuznetzova 1928).

The temperature dependency of universal R2 in wheat is shown in Figure 10.4c by the combined information from Gabo and Prelude varieties. One may therefore expect that if a variety is transferred to a cooler environment, the length of its third stage may be nearly doubled as with the change from a mean temperature of 20° to 13°C (from about 35 to 65 days).

The pattern of development between emergence and heading in two of the varieties indicates the presence of positive responses to photoperiod. In Figure 10.4b the Prelude wheat is compared when grown for five seasons under continuous light at high latitude (summer sown at Khibiny 67°N) and under shorter photoperiod at lower latitude (spring sown at Dnepropetrovsk 48°N). The temperature-dependent R1 reduces the period with increased temperature at both places. In addition, sensitivity to photoperiod shortens the period at the higher latitude as shown by a reduction of 40 per cent when the variety grew at both places at 17°C. The variety is likely to be further retarded if grown at still lower photoperiod.

This expectation is confirmed by the behaviour of a similar variety, Thatcher wheat, when grown in different latitudes. Figure 10.4a reveals the action of the same two types of response in a wider range of photoperiod, from continuous light in Alaska to 12 hours per day at the sites in Mexico and Hawaii and to 10 hours per day when winter-sown at Melbourne. The response is much greater as shown by the fact that heading is reached in one-third the time (reduced from 115 to 40 days) when grown at about 11°C in Alaska compared with Mexico.

Further analysis is also possible, because the date of FI is available from these sites. Figure 10.4a shows that both R5 and R6 in Thatcher wheat contribute to the length of the period from emergence to heading. If the sowing at site 2 in Mexico is again compared with that in Alaska at 11°C, it can be

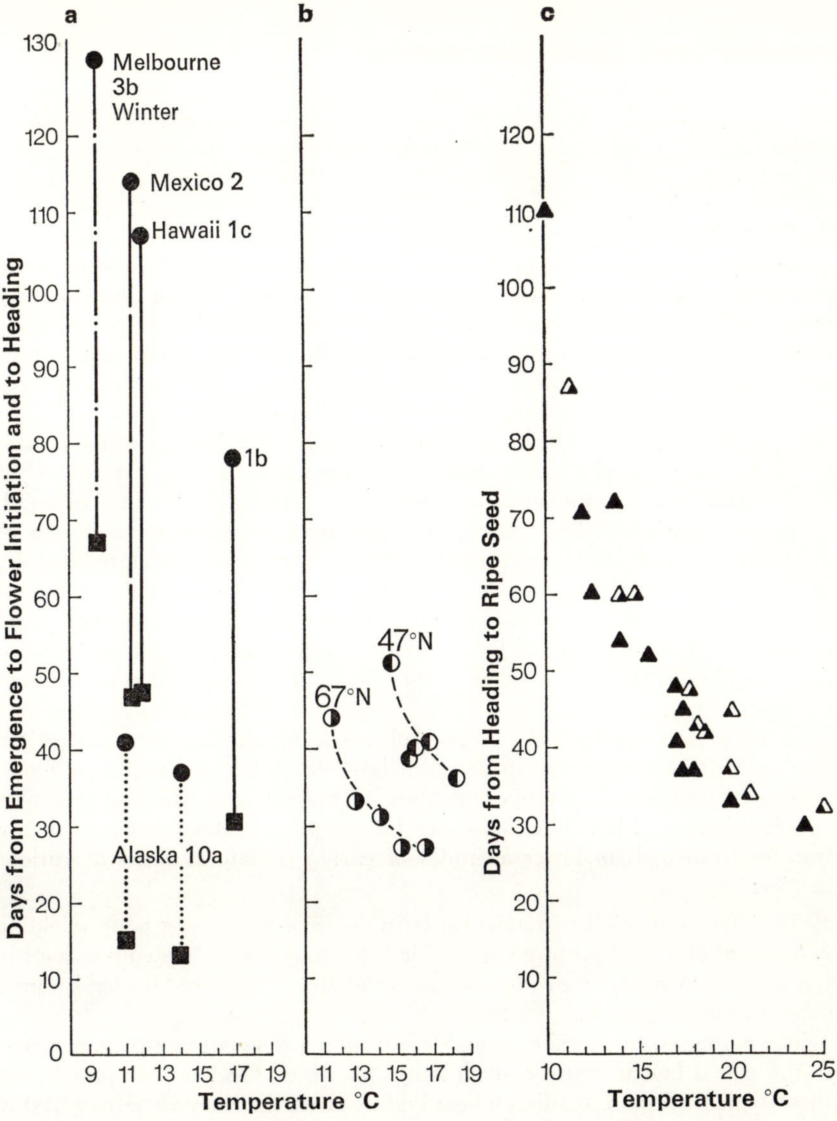

Fig. 10.4 Temperature is related to stage of development in wheat according to the type of responses in the MG, as shown in the varieties Gabo and Thatcher grown at the climate sites, and in Prelude grown in Russia. (a) In Thatcher wheat with its photoperiod-positive responses the trend of the daily mean temperature with fewer days from emergence to FI (square) and to heading (circle) is complicated by the hastening effect of longer photoperiod on rate of development in both stages. Plants grown in a shorter photoperiod (sites 3b winter, 2, 1bc) are much slower than those grown in continuous light (site 10a, 1965 and 1970) but in similar temperature. (b) Because Prelude wheat has similar responses, data from sites at two latitudes over 5 years of testing show the same trend of temperature with fewer days from emergence to heading but at a lower level on the graph for the higher latitude. The period is shorter at the same temperature in plants grown in continuous light at Khibiny 67°N, sown in early summer, than in a shorter photoperiod at Dnepropetrovsk 48°N, sown in spring. (c) The trend of mean temperature with fewer days from heading to ripe seed through universal R2 coincides in the two early wheats, Gabo and Prelude. Gabo, solid triangle; Prelude 5 years at Kotlas 61°N, triangle with right corner solid; Prelude 5 years at Dnepropetrovsk, triangle with left corner solid.

seen that both responses are strong. R5 reduces the first stage to one-third (from 47 to 15 days) and R6 reduces the second stage almost as much (from 68 to 27 days).

Examples of selection

An important guide-line in selection is that certain kinds of MG are more suited than others to a given environment. This fact may be illustrated by the following examples.

Growing period and altitude. Varieties may be needed for growing seasons of the same length but different temperature, as may occur with change of altitude at the same latitude (chapter 6). This problem can be solved if varieties of category 2 are available (Table 9.3). The MG must have sufficiently intense individual responses to compensate for the interaction of temperature with the universal responses. A variety with a particular length of growing period at low temperature can only have the same length at higher temperature if temperature-negative responses of moderate intensity are present. These must delay development just sufficiently to balance the increased rates due to the operation of R1 and R2, as in early Cape barley and late Algerian oat (Fig. 5.6). When the variety is transferred from high to lower temperatures, the same responses adjust the rate of development in the reverse way.

Growing period and latitude. The problem is more difficult if varieties are needed for growing seasons in different latitudes. The seasons may differ in both temperature and photoperiod, thus strongly affecting the expression of certain kinds of MG. The following two examples consider the effect of transfer from high to lower altitudes of early and late varieties of various species:

1. Transfer from high to moderate latitude. The search for better-yielding varieties of crop and pasture species for the winter-spring growing seasons in the temperate zone, often includes material from the even shorter summer seasons at high latitude.

The winter sowing at site 3b at Melbourne represents the growing seasons in the wheat belt in south-eastern Australia which are long compared with those typical of the Canadian wheat belt as at site 6. They vary from 160 to 200 days compared with about 80 days and have lower values of temperature and photoperiod as shown in Table 8.1. Because the growing season is actually short in both types of climate, varieties grown for seed must reach FI at a low VN to allow adequate flowering and ripening of seed. The change to FI must occur before about 30 days after sowing at the higher latitude and 70 days after sowing at the lower one.

The most important fact affecting development at Melbourne is that the universal responses are slowed by the cooler season, particularly in the first three months which are close to 10°C. In addition, although temperature-negative responses are not activated because of the low temperature, the short photoperiod after sowing strongly retards FI in varieties with R5.

Thus suitable varieties for the short winter-spring season must have in common a low minimum VN and be little retarded by individual responses.

Varieties in category 1 (Table 9.3) give the most rapid development because of their simple MG. Others in categories 3 and 4 may also be useful as long as the photoperiod responses are only slight, because the temperature-negative responses are unexpressed. The varieties include early peas, subterranean clovers and cereals.

In contrast the short summer season in Canada permits a wider range of MG because those with a strong R5 reach FI rapidly in the long photoperiod, as with the later peas and both early and late red clovers. Figure 9.6 shows the long growing period of two barley varieties in the Melbourne winter compared with the Canadian summer. Bankuti barley (category 1) takes 170 days, as shown by the left column of the pair, while Cape barley (category 4) takes 200 days, as shown by the middle column of the group of three. Both varieties are much faster in Canada (about 80 days) as shown by the right-hand columns. Even the late Mackay pea and Montgomery red clover develop remarkably fast as shown by the right member of each pair of examples. They take only 80 and 90 days respectively because of their positive responses to the high values of both photoperiod and temperature.

2. Transfer from moderate to low latitude. With diversification of tropical agriculture, it is important to understand not only the tropical species but also the change in development that may occur in varieties of temperate species when transferred to the tropics. The following example compares some of the varieties grown at site 3b in the winter-spring season at Melbourne with their development after sowing in winter at site 1a in Hawaii (chapter 6). The temperature is much higher (+13°C) and the photoperiod slightly longer (+2 hr/dy) as discussed also for sites in northern Australia (p. 121).

In the cool season at Melbourne early pea and barley (category 1) take 150 and 170 days respectively as shown by the left-hand columns in Figure 9.6. Early subterranean clover (Table 8.1) is similar to the barley and the late one takes about 200 days (see the left-hand column for the Tallarook variety in Fig. 9.6). In the hot tropical season, however, the period varies greatly according to MG. The varieties in the first category take only half the time, but both varieties of subterranean clover (category 3) are much delayed (see right-hand column for each variety in Fig. 9.6).

In the early variety of subterranean clover the high temperature activates R3 and lengthens the first stage at least three-fold (from 27 days to about 90 or more). However, the other two stages are shorter than at Melbourne because of the universal responses to the high temperature. Nevertheless the plants are so delayed in FI and erratic in flowering that they are strikingly transformed in habit from their usual sparse-leafed early form to a very leafy late one as at Ayr (Plate 4B). In the late variety, a stronger R3 prevents FI entirely.

New combinations

Although the knowledge of certain kinds of MG is at present limited to one or two species, the use to which they can be put suggests the value of definite searching to see if they occur in other species.

For instance, the combination of low VN and the universal temperature responses only or with slight negative ones, provides varieties with comparatively short growing periods, irrespective of temperature, in both pea, oat and wheat. This combination, originally selected unwittingly for the wheat belt in southern Australia, has enabled oat and wheat to be grown successfully in the tropical part of the continent (Beech 1966ab). It is also the basis for Carnamah, the very early variety of subterranean clover, which has extended the range of the species into the short growing season of the dry edge of the wheat belt in Western Australia (Taylor 1967).

A new combination needed for short, cool growing seasons is one with a low VN and with the universal temperature responses only, a faster rate of leaf appearance than pea and a fast maturation rate at low temperature.

Seed production

Nowadays varieties of agricultural species are often grown for seed as a specialized enterprise, even in climates outside the area to which they are adapted for use. The choice of location and the management of the seed crop must be based on maintaining trueness to the MG as well as on the prospect of a high seed yield per hectare. Otherwise a 'shift' of MG may occur to a population either earlier or later than the original one (Evans 1960). The change is more likely to occur within varieties of cross-pollinated species than of self-pollinated ones, as recorded for Kent perennial ryegrass (Cooper 1959b), Pilgrim white clover (Stanford 1962) and Tammisto red clover (Garrison 1961).

Genetic change is prevented by growing the crop where the desired proportion of the variable population will be the most distinct, and managing it to this end. The crop may also need to be limited in number of harvests in order to prevent unwanted variation through recombination within the population or from cross-pollination from local different types. The red clover species provides a good example.

Development at high latitudes. The management of late-flowering (single-cut) varieties of red clover for seed varies somewhat according to the latitude. When grown in high latitude, the crops are sown in spring but harvested after the following winter for seed—that is, in their second flowering season.

In the first season, both early and late varieties develop fast after sowing in long photoperiod and flower together in that summer, as in Canada and Wales, sites 6 and 7, (see Fig. 9.6, Table 6.5). However, the reproductive lateral shoots on each plant are followed by others which remain vegetative much longer because of the lowering photoperiod. In the following spring the increasing photoperiod reveals the range of MG within the population because the plants with a lower threshold of response reach FI earlier than those with higher thresholds.

In Wales an early MG like Cowgrass reaches FI in early spring and begins to flower in early summer while the late Montgomery variety is much slower and most of the plants begin to flower about a month later. Since the early-flowering types are more susceptible to cold than the late ones (Therrien 1960), their presence in the crop may be eliminated before the next spring if the first winter is harsh enough. If not, the survivors can be identified by their earlier flowering and the seed grower can thus estimate the value of the crop for seed of the desired late variety.

Development at lower latitudes. Where seed is produced in lower latitudes such as in New Zealand, the milder winter precludes much loss of the earlier-flowering types after sowing the previous spring or summer. However, differences in intensity of R5 within the seed crop will increase the range of flowering in the following spring and summer even more than at higher latitudes. This occurs because plants in New Zealand grow to some extent through the winter and the photoperiod is shorter in spring than at higher latitudes.

Their development to flowering is likely to follow the same pattern as in the similar climate at Melbourne. Here the Cowgrass variety reaches FI from mid-winter onwards and begins to flower from mid-spring, while the late Montgomery only begins FI at about this time and reaches first flower in early summer, about $2\frac{1}{2}$ months after the early one. This large difference can be used to assess the trueness to type of the seed crop even though each variety varies greatly between plants in the time of first flower.

Planned adjustment of MG. For many centuries, crops have been improved genetically by selecting superior plants at harvest time and carefully increasing them. Pusey (1842) records a new, early-maturing oat variety from Scotland which was 'raised from a single ear by Mr. Dyock near Aberdeen in 1830' and 'of value where chill summers seldom ripen the crop'. Nowadays increased knowledge of the maturity character allows one to understand the complex basis to the simple fact of a few days' gain in seed maturity, as indicated in wheat varieties (Table 10.3).

In consequence 'controlled climate' tests, such as for sensitivity to photoperiod (Allard 1932), can be used to reveal desired genetic variation within or between populations, preferably at an earlier stage than that of ripe seed. A population that includes plants with different intensities of the one type of individual response will be most variable when growing at the threshold value of the important climatic factor. This will occur in lower photoperiod for the varieties with R5 and R6 or in higher temperature for those with R3 and R4. Thus early-flowering plants for either type of environment may be the more easily selected for seed. This approach was used to identify winter wheats able to head in moderate photoperiod at high temperature (Blair 1962) and thus able to seed in a wide range of climate in the United States. The method can also be applied to the study of genetic variation within and between stages, such as the vegetative compared with the early reproductive stage in cereals.

DEVELOPMENT AND YIELD

The relation of development to yield is 'the central problem of plant science' (Whyte 1960). A high yield of dry matter is associated with fast growth rate and slow development. Because both total and seed yield of annual crops tend to increase with longer growing period (Bell 1966, Gladstones 1971), varieties need to be selected to fit the growing season. A shorter growing period may lower yield by limiting the intake of solar energy (van Dobben 1962) or by increasing cold injury due to precocious development. This occurs with the early cereals when sown in autumn in Wales or in late summer in southern Australia (Table 8.1, Fig. 8.2). Unfavourable weather may also lower yield in varieties with a growing period longer than the growing season. Thus drought may reduce the seed setting and filling of late cereals in Australia (Aitken 1963).

Trends of yield and development

The general trend of higher yield with slower development raises an important question. Where varieties of a crop are already well suited in growing period to the season, can the rate of development be altered to give a higher yield of seed? Delayed FI certainly increases the number of reproductive nodes on the young inflorescence of cereals (as with Research barley in Plate 4A). This delay thus contributes in Europe towards the trend to higher grain yield in winter wheats, compared with spring varieties. However, delayed FI usually retards flowering and seed ripening.

Table 10.3 gives a local illustration of the effect of development on yield from early wheats sown in winter and grown in normal or in continuous photoperiod. Both the Gabo and Thatcher varieties develop more slowly in normal photoperiod as shown by higher VN and by longer periods from sowing to FI, heading and ripe seed.

The other information indicates that the yield is also higher, as measured

Table 10.3
The effect of photoperiod on development rate and on yield and allied measures in two early wheats at Derrimut near site 3b, sown winter 1972*

Photoperiod hr/dy	24	natural†	24	natural
Measures on main shoot	Gabo wheat		Thatcher wheat	
Development rate				
stage 1 (days)	31	45	34	60
stages 1 + 2 (days)	66	88	68	117
growing period (days)	131	146	133	166
vegetative nodes‡	5·4	7·4	6·0	10·5
Yield				
seed—number	14	30	12	29
—weight (g)	0·6	1·3	0·5	0·9
Yield characters				
reproductive nodes‡				
total	9·7	14·7	10·8	19·3
fertile	7·7	13·7	4·7	15·5
seeds/fertile node	1·9	2·2	2·6	1·9
seed size g/100	4·4	4·3	2·5	3·2

* temperature °C, mean daily each month after sowing: 8·5, 10·1, 12·8, 13·2, 14·6, 17·5
† mean monthly after sowing: 11·0, 11·9, 13·0
‡ each mean value is from 6 plants and has a mean coefficient of variation of 3%

by the seed number and weight, both of which are doubled. Inspection of the four yield characters shows that the increased yield is based almost entirely on only two of them, the total and the fertile reproductive nodes. Both these measures are related to the time of FI.

One may speculate that an MG might be possible that produces the same length of growing period as Gabo wheat but by somewhat different units. The MG might combine a slightly slower tendency to flower than Gabo, with faster rates of development after FI, through appropriate responses. This type of MG would increase the VN, the number of leaves and the reproductive nodes per shoot as in the later-maturing varieties. However, faster development after FI might improve reproductive efficiency and thus the weight of seed if combined with other yield characters such as high seed setting per spikelet.

Stability of yield in different climates

Another question concerns the stability of high-yielding varieties of a crop when transferred to other types of growing season (Finlay 1963, Stakman 1967). This is a matter of great practical significance in large-scale schemes for higher crop yields in needy countries and is also of historical interest in Australia. Given that plant nutrition, disease and input of solar energy are not limiting, the problem of stability is concerned with the interaction between the MG and the new type of growing season. The introduced variety may develop too slowly or too fast, thus reducing its actual yield. In the first

Table 10.4
The source of wheat varieties affects their development rate, sown winter at site 2 in Chapingo, Mexico, 1970

Measures of development*	Source of variety groups†		
	Low latitude Mexico	Moderate latitude Australia	High latitude North America Europe
1. Stage 1 (days)	41	43	51
2. Stage 2 (days)	55	55	73
3. Stage 3 (days)	56	53	43
4. Growing period (days)	151	151	167
5. Difference in days compared with Mexico	—	0	+16
6. Leaf stage at FI	5·4	5·4	6·7
7. Vegetative nodes on shoot	10·5	10·1	12·7

* coefficient of variation usually less than 2% ; all values for the high latitude wheats are significantly different from the other two groups

† seven early varieties selected from each of three sources
1. Mexico: Ciano, Inia 66, Nuri, Pitic, Potam, Saric, Yecora
2. Australia: Gabo, Insignia, Mendos, Mengavi, Spica, Timgalen, Triple Dirk
3. North America and Europe: Gasser, Kogat (Alaska); Selkirk, Thatcher (Canada); Kolibri, Opal, Tammi (Europe)

case the inflorescences may have high yield potential but fail to set much seed. In the second case the inflorescences may each only form a few reproductive nodes and thus limit the number of grains matured.

This can be demonstrated by reference to the nature of the MG in the early cereal varieties used for short growing seasons in lower latitudes compared with higher ones. Varieties suited to the spring-summer season of the wheat belt in Canada have a growing period of about 80 days as shown at site 6 for Cape barley (Fig. 9.6). They develop fast because of their low minimum VN and the effect of high temperature on R1 and R2. Strong photoperiod responses may also be present but are not selected for or against because these varieties develop as fast in the long photoperiod as insensitive ones. Thus many varieties such as Thatcher wheat have a strong R5 and R6 (Table 10.4), although others like Edda and Olli barley have only weak responses.

In the winter-spring growing season of lower latitudes, the selection pressure is different. Suitable varieties must have a growing period close to 180 days in the wheat belt of southern Australia (site 3b) or about 150 days as an irrigated crop in the highlands of Mexico (site 2). The varieties must develop fast in moderate photoperiod and at more or less low temperature. They must therefore have been selected not only for low minimum VN but also for low sensitivity to photoperiod in both the first and second stages, as shown in Australian wheats such as Gabo and in Mexican wheats such as Ciano. Thatcher wheat, grown alongside these varieties in their respective usual growing seasons, takes several weeks longer to reach ripe seed because flowering is delayed by the moderate photoperiod. This delay can be crucial to seed yield where droughts shorten the growing season.

Table 10.4 illustrates the patterns of development in three groups of seven varieties in relation to their source, when grown at site 2 in Mexico. The samples indicate that the Mexican and the Australian wheat breeders have achieved the same general MG by selecting for short growing period in their respective types of climate. These groups of varieties are alike in each of the six measures of development but different from the high latitude group, which has a longer vegetative stage. In this group each variety reaches FI more slowly because of a stronger R5 and therefore has a higher leaf stage at FI and a higher VN at heading. The second stage is also slower because each variety has a stronger R6. However, as the heading is delayed until mid-spring, the third stage is shorter because of the universal response R2 to the higher temperature and because of moisture stress. Thus the stronger R5 and R6 in the MG of this group retard the time of heading by 27 days (9 days in the first stage and 18 in the second) but the faster rate of seed ripening reduces the difference in growing period to 16 days.

3 Further Aspects of Development Analysis

UNDERSTANDING THE PHENOTYPE

Development analysis sheds light on hitherto puzzling plant reactions to different environments. This is illustrated by two aspects of cereals at high latitudes. Albright (1933) noted the rapid growth in early summer of wheat and oat crops at Fort Vermilion, Canada (58°N), compared with the same varieties sown at the same time at Beaverlodge, 360 kilometres farther south. The present information from climate sites 6 and 10a at 52° and 65°N respectively, indicates that the taller crops at Fort Vermilion are likely to have resulted from earlier FI in the longer photoperiod of that growing season and hence earlier elongation of the shoots.

Another aspect is the different relationship of development to temperature before and after heading. Carder (1957) examined the same varieties of wheat, barley and oat at Madison, Wisconsin (43°N) and at Beaverlodge, Alberta (55°N), which is cooler but has a photoperiod 4 hours longer in the first month after sowing. In all cases emergence to heading was about the same at the two places but sowing to emergence and heading to ripening were much slower at Beaverlodge. Carder put forward the view that the unexpectedly fast development up to heading resulted from a compensating interaction of the longer day with the lower 'effective heat'. Information from the spring-sown climate sites from moderate to high latitude suggests that the longer photoperiod at Beaverlodge would indeed hasten varieties with R5 and R6 by lowering VN and increasing the rate of leaf appearance after FI. In addition the temperature close to the ground could be a little higher than in the standard screen, thus also hastening early development (pp. 53–5).

FORECASTING CROP HARVESTS

Variety

The concept of development analysis clarifies the problem of forecasting the time of harvest of a crop (pp. 157–60). In consequence, when the one variety is to be grown in a series of sowings at the one place, varieties of complex MG will be avoided if possible. In varieties of simple MG, temperature will be the positive controlling factor through the universal responses, for the period from sowing to harvest.

When a crop is grown through a wide range of latitude, even a simple MG may be a little more closely correlated with the multiple of temperature and photoperiod than with temperature alone (Table 10.1). If the varieties are of more complex MG, the individual responses must be identified and measured in order to anticipate their effect on rate of development.

Varieties with strong responses to photoperiod can be expected to affect the length of the period up to first flower when grown in lower latitudes compared with high ones. With Thatcher wheat, for instance, the period to heading is lengthened at lower latitude. Thus the whole growing period may be lengthened because of the strong R5 and R6 (Fig. 10.4a, Table 10.3). In other species such as proso millet, varieties with photoperiod-negative responses affect development in the opposite way (Kuznetzova 1928).

In a promising variety the knowledge of its MG not only permits the forecasting of its pattern of development according to growing season but provides the basis for a more critical evaluation. The variety should be assessed for characters that may arrest development through lethal sensitivity to environmental stresses. These characters include the susceptibility of reproductive tissues to unfavourable temperatures and water supply. Frost may damage both the elongating shoot in cereals and the flowering head. In grain sorghum, even a brief exposure to moderate temperature (13°C) just before heading kills the pollen mother cells and thus causes male sterility (Downes 1971). In the Washington Navel orange daily water deficits may cause the young fruit to dehisce (Nuttonson 1959). Once the character is identified, improvement of the variety may be sought, if possible by genetic means, as in the programme for faster-ripening subterranean clover (Tennant 1969) in Western Australia. Resistance to such stresses may be of key importance to the success of a crop along the margin of the most suitable zone for its cultivation.

Measures for forecasting

Analysis is needed to judge the situation in which other measures may be preferable to thermal units based on mean daily temperature. A multiple of temperature and solar radiation (Caprio 1971) or of temperature and photoperiod (Nuttonson 1948, 1957) may be more closely related to the universal responses to temperature of the low-growing plant in growing seasons with high daily heat loss. This situation occurs at high latitudes in summer or at high altitudes in the tropics in winter.

Nuttonson records a trend in the length of growing period of a number of crops that is related more to the sum of the photothermal units than to the total heat units of each growing season. The total heat units were fewer with longer photoperiod both when the same variety was sown in a range of latitude or from spring to summer at the same place. The total photothermal units remained about the same over the range of latitude or season.

Since the trend was found in both the Alaska pea, which is insensitive to photoperiod, and in Thatcher wheat, which is sensitive, the evidence implicates the universal responses in the simple MG of the pea and both the universal and individual responses in the more complex MG of the wheat.

A more complex measure, the 'development unit', is related to potential evapotranspiration, solar radiation and rate of pea node appearance (Thornthwaite 1954). This measure is used in the complex management of a series of irrigated vegetable crops in New York State. The positive relationship of the rate of pea node (or leaf) appearance with temperature assists the planning of dates of sowing and harvest. In addition wilting may slow both growth and development of the crop. Since the rate of node appearance is a component of both processes, attention to the rate from day to day in the crop can enable the agronomist to plan a more precise irrigation system and thus forestall water stress.

THE PRESENT CHALLENGE

As plants are grown and processed in a more complex way, both the farmer and the processor need to understand more clearly the factors that control the development of their particular crop from sowing to harvest. Only in a few cases so far, notably in horticulture, is the knowledge sufficient.

Thus in the chrysanthemum many varieties are of the 'short day' type, and in gardens in temperate latitudes flowering usually begins in autumn. Nowadays, however, the city dweller can buy the plant for home decoration at other times of the year because suitable varieties develop fast under controlled temperature and photoperiod.

In most crop and pasture species the essential knowledge is still missing. In each species the scattered information on MG in various countries should be integrated by the observation of a range of identical genotypes in a wide diversity of climates as well as in controlled environments. The genotype-climate studies reported in this book provide the principles needed for the effort and have begun to span the gap of ignorance in a few species.

Appendix 1

Common and Latin names of species mentioned in the xt

Apple	*Pyrus malus* L.
Barley	*Hordeum vulgare* L.
Beans—tropical	*Phaseolus* spp. (Tourn.) L.
Beet	*Beta vulgaris* L.
Cauliflower	*Brassica oleracea* L.
Celery	*Apium graveolens* L.
Chrysanthemum	*Chrysanthemum morifolium* Hemsl.
Chick-pea	*Cicer arietinum* L.
Clover—red	*Trifolium pratense* L.
Clover—subterranean	*Trifolium subterraneum* L.
Clover—white	*Trifolium repens* L.
Cocksfoot (orchard grass)	*Dactylis glomerata* L.
Cotton	*Gossypium hirsutum* L.
Crocus	*Colchicum* sp.
Cucurbits	*Cucurbita* spp. (Tourn.) L.
Daffodil	*Narcissus pseudo-narcissus* L.
Dandelion	*Taraxacum officinale* Weber
——	*Dupontia fischeri* R. Br.
Fig	*Ficus carica* L.
Flax	*Linum usitatissimum* L.
Grape	*Vitis vinifera* L.
Hemp	*Cannabis sativa* L.
Hops—Japanese	*Humulus japonicus* Sieb.
Maize	*Zea mays* L.
Millet—common, proso, Indian	*Panicum miliaceum* L.
Oak	*Quercus* sp.
Oat—common, red	*Avena sativa* L., *byzantina* C. Koch
Orange	*Citrus aurantium* L.
Paspalum (dallis grass)	*Paspalum dilatatum* Poir.
Pea	*Pisum sativum* L.
Plum	*Prunus domestica* L.
Potentilla	*Potentilla glandulosa* Lindl.
Radish	*Raphanus sativus* L.
——	*Ranunculus nivalis* L.
Rape	*Brassica napus* L.
Rice	*Oryza sativa* L.
Rye (ryecorn)	*Secale cereale* L.
Ryegrass—perennial	*Lolium perenne* L.
Ryegrass—annual	*Lolium rigidum* Gaud.
Snowdrop	*Galanthus nivalis* L.
Sorghum	*Sorghum vulgare* Pers., *bicolor* (L.) Moench
Soybean	*Glycine max* Merrill
Timothy grass	*Phleum pratense* L.
Tobacco	*Nicotiana tabacum* L.
Wheat—common or bread	*Triticum aestivum* L.
Wheat—emmer	*Triticum dicoccum* Schubl.,
Wheat—einkorn	*Triticum monococcum* L.

Appendix 2

The relationship of the experimental site and time of sowing to the mean daily temperature, photoperiod (including civil twilight) and solar (global) radiation of the first month after sowing

Sites	Times of sowing		Mean daily values for first month Temperature °C	Photoperiod hr/dy	Radiation cal/cm²/dy
1a	winter	21 Feb 1963	23·2	12·6	460
1b	winter	5 Feb 1963	16·9	12·4	470
1c	winter	6 Feb 1963	11·8	12·4	520
2	winter	17 Dec 1970	10·3	11·8	450
3b	winter	30 May 1963	9·9	10·6	150
3b	spring	25 Oct 1963	16·0	15·3	500
3b	summer	17 Feb 1964	17·0	14·0	450
4a	spring	5 Apr 1963	12·1	14·4	490
4b	summer	8 June 1963	14·8	15·8	700
4c	summer	20 July 1963	11·0	14·7	720
5	spring	26 Apr 1963	12·1	15·8	540
6	spring	17 May 1963	13·0	17·6	490
7	spring	21 May 1963	14·6	17·8	380
7	autumn	5 Sept 1963	12·7	12·8	190
8	summer	7 Dec 1968	6·3	18·8	460
9	summer	2 Dec 1970	4·9	18·8	520
10a	spring	18 May 1965	8·1	24·0	530
10a	spring	29 May 1970	12·5	24·0	490
10b	summer	5 July 1970	3·2	24·0	510

Appendix 3

Solar (global) radiation at experimental sites or equivalents (monthly mean in cal/cm²/dy)*

Site	Year	Jan	Feb	Mar	Apr	May	June	July	Aug	Sept	Oct	Nov	Dec
1a Hawaii	1963	370	430	470	490	550	550	500	480	450	420	390	360
1b Hawaii	1963	420	460	510	520	580	610	570	540	530	460	410	390
1c Hawaii	1963	440	510	560	550	620	670	600	600	560	490	430	410
2 Mexico	1971	450	530	530	590	550	510	500	490	440	490	410	440
3b Australia	1963	560	500	410	300	170	150	180	260	250	400	520	550
	1964	580	450	450	250	220	160	160	210	330	400	480	510
	1965	600	520	450	320	200	170	170	240	330	460	500	540
	1968	660	560	380	280	140	140	160	250	330	390	510	580
	1969	640	450	400	280	180	160	170	210	310	430	530	570
4a California	1963	190	300	410	480	600	760	760	660	490	370	190	110
	1964	190	—	450	—	—	670	720	660	500	340	200	150
4b California	1963	200	300	400	400	500	700	700	650	430	370	190	100
	1964	190	300	400	400	500	650	710	640	—	—	—	—
4c California	1963	—	—	—	—	—	—	750	700	520	—	—	—
5 Oregon	1963	100	220	290	430	500	540	580	500	350	230	120	70
6 Canada	1963	100	170	290	410	480	506	590	460	320	220	110	80
	1964	110	210	370	420	500	560	560	430	250	200	90	60
7 Wales	1963	50	90	170	270	350	390	360	360	200	150	70	40
	1964	50	90	190	270	330	390	350	300	180	150	70	40
8 Macquarie I.	1968	390	330	190	120	50	30	50	70	140	170	330	400
	1969	350	280	200	100	50	20	40	50	170	230	300	—
9 South Georgia I.	1970	—	—	—	—	—	—	—	—	—	—	—	520
	1971	430	380	260	—	—	—	—	—	—	—	—	—
10a Alaska	1965	20	90	170	300	500	560	490	360	200	100	20	10
	1970	10	80	180	420	510	520	400	370	280	100	30	10
10b Alaska	1970	2	30	170	350†	540†	550	520	230	110	40	5	0

* 1 cal/cm² = 1 langley; 1 cal = 1·163 mw hr = 4·184 joules
† mean value 1964-8

Sources: Sites 1, 4, 5, 10, Department of Commerce, U.S.A.; Site 2, Meteorological station, Escuela National de Agricultura, Chapingo, Mexico; Site 3b, Bureau of Meteorology, Australia; Site 6, Canadian Department of Transport; Site 7, U.K. Meteorological Office and Welsh Plant Breeding Station, Aberystwyth; Site 8, Antarctic Division, Department of External Affairs, Melbourne, Australia; Site 9, British Antarctic Survey, Birmingham, England.

Appendix 4

A classification of growing seasons in relation to three measures of temperature and one of photoperiod (including civil twilight) over the period of 90 days from sowing

Experimental sites		Time of sowing	Type of season	Mean daily values for 90 days from sowing: Temperature °C mean	max.	min.	Photoperiod hr/dy
10b*	Alaska	summer	cool	2·5	4·8	0·1	23·0
8	Macquarie I.	summer		7·0	8·6	5·4	18·0
9	South Georgia I.	summer		5·2	8·2	2·5	18·0
7	Wales	autumn		10·6	14·3	7·8	11·8
3b	Australia	winter		9·0	12·5	5·7	10·8
4c*	California	summer		11·0	19·0	4·8	14·7
1c	Hawaii	winter		11·8	16·2	7·3	12·7
10a	Alaska	spring 1965		12·8	19·3	6·8	24·0
10a	Alaska	spring 1970		13·9	20·7	7·1	22·2
2	Mexico	winter		11·1	23·7	−1·3	12·2
7	Wales	spring	warm	14·5	18·5	10·8	17·8
5	Oregon	spring		14·0	20·3	8·0	16·3
4a	California	spring		15·7	20·7	9·5	14·7
3b	Australia	summer		16·4	20·7	10·9	12·8
4b	California	summer		17·1	25·7	8·5	15·6
3b	Australia	spring		17·2	22·1	12·3	15·6
1b	Hawaii	winter		17·0	21·7	12·3	12·8
6	Canada	spring		17·2	23·5	10·9	17·5
1a	Hawaii	winter	hot	23·5	26·0	21·0	13·1

* Mean values are for the 50 days from sowing to harvest. Ground free from snow July to September. No plants reached flowering.

References

Abbe, C. (1905). A First report on the relations between climates and crops. *Bull. Weath. Bur. U.S. Dep. Agric.* no. 36. Washington, D.C.

Abbe, E. C., Randolph, L. F. and Einset, J. (1941). The developmental relationship between shoot apex and growth pattern of leaf blade in diploid maize. *Am. J. Bot.* **28**: 778-84.

Agrawala, V. S. (1963). *India as Known to Panini.* 2nd ed. Benaras Hindu Uni.

Aitken, Y. (1955a). Flower initiation in pasture legumes. I. Factors affecting flower initiation in *Trifolium subterraneum* L. *Aust. J. agric. Res.* **6**: 212-44.

—— (1955b). Flower initiation in pasture legumes. II. Geographical implications of cold temperature requirements of varieties of *Trifolium subterraneum* L. *Aust. J. agric. Res.* **6**: 247-57.

—— (1955c). Flower initiation in pasture legumes. III. Flower initiation in *Medicago tribuloides* Desr. and other annual medics. *Aust. J. agric. Res.* **6**: 258-64.

—— (1961a). Flower initiation in relation to maturity in crop plants. I. Flower initiation in relation to maturity in some Australian oats (*Avena* species). *Aust. J. agric. Res.* **12**: 371-88.

—— (1961b). Flower initiation in relation to maturity in crop plants. II. Internode lengthening in relation to grazing in oat varieties. *Aust. J. agric. Res.* **12**: 389-96.

—— (1962). Shoot apex accessibility and pasture management. *J. Aust. Inst. agric. Sci.* **27**: 50-2.

—— (1963). The flowering character of the first wheat grown in Australia. *Aust. J. Sci.* **25**: 496-7.

—— (1964). Flower initiation in pasture legumes. IV. Flower initiation in *Trifolium pratense* L. *Aust. J. agric. Res.* **15**: 21-36.

—— (1966a). Flower initiation in relation to maturity in crop plants. III. The flowering response of early and late cereal varieties to Australian environments. *Aust. J. agric. Res.* **17**: 1-15.

—— (1966b). The following responses of crop and pasture species in Australia. I. Factors affecting development in the field of *Lolium* species. *Aust. J. agric. Res.* **17**: 821-9.

—— (1967). Leaf primordia formation in some agricultural species. *J. Aust. Inst. agric. Sci.* **33**: 212-14.

—— (1970). Non-destructive methods for estimation of flower initiation in subterranean clover and cereals. *J. Aust. Inst. agric. Sci.* **37**: 57-60.

—— and Drake, F. R. (1941). Studies on the varieties of subterranean clover. *Proc. R. Soc. Vict.* (N.S.) **53**: 342-93.

—— and Haughton, J. M. (1945). The species *Pisum sativum* in relation to Australian agriculture. *J. Aust. Inst. agric. Sci.* **11**: 35-40.

Albright, W. D. (1933). Crop growth in high latitudes. *Geogrl Rev.* **23**: 608-20.

Allard, H. A. (1932). Length of day in relation to the natural and artificial distribution of plants. *Ecology.* **13**: 221-34.

Anati, E. (1963). *Palestine before the Hebrews.* London.

Azzi, G. (1956). *Agricultural Ecology.* London.

Beech, D. F. and Norman, M. J. T. (1966a). The effect of time of planting on yield attributes of wheat varieties in the Ord river valley. *Aust. J. exp. Agric. Anim. Husb.* **6**: 183-92.

—— (1966b). The effect of time of flowering on yield attributes of oat varieties in the Ord river valley. *Aust. J. exp. Agric. Anim. Husb.* **6**: 250-4.

—— (1967). The effect of time of planting on the performance of irrigated winter crops in the Ord river valley. *J. Aust. Inst. agric. Sci.* **33**: 77-84.

Bell, G. D. H. and Kirby, E. J. M. (1966). Utilization of growth responses in breeding new varieties of cereals, in F. L. Milthorpe and I. D. Ivins (eds), *The Growth of Cereals and Grasses*, p. 308. London.

References

Bigelow, J. (1818). Facts serving to show the comparative forwardness of the spring season in different parts of the United States. *Mem. Am. Acad. Arts Sci.* **4** (1): 77-85. Excerpts with comments, *Am. J. Sci.* **1**: 76-7.

Blackman, G. E. (1968). The application of the concepts of growth analysis to the assessment of productivity, in F. E. Eckardt (ed.), *Functioning of Terrestrial Ecosystems at the Primary Production Level.* UNESCO Proc. Copenhagen Symp., Paris.

Blair, B. O. and Patterson, F. L. (1962). Controlled climate tests for evaluating the adaptation of winter wheats. *Crop Sci.* **2**: 374-7.

Bliss, L. C. (1956). A comparison of plant development in micro-environments of arctic and alpine tundras. *Ecol. Monogr.* **26**: 314-37.

Bomalaski, H. H. (1948). Growing degree days. *Fd Pckr.* **29**: 51-61.

Bonnett, O. T. (1935). The development of the barley spike. *J. agric. Res.* **51**: 451-7.

Bradfield, R. (1969). Training agronomists for increasing food production in the humid tropics, in *International Agronomy: training and education.* Am. Soc. Agron. special publ. no. 15.

Braidwood, R. J. (1960). The agricultural revolution. *Scient. Am.* **203**: 131-48.

Britten, E. J. (1961). The influence of genotype and temperature on flowering in *Trifolium repens. Agron. J.* **53**: 11-14.

Brown, J. and West, G. C. (1970). Tundra Biome Research in Alaska: the structure and function of cold-dominated ecosystems. *U.S. IBP-Tundra Biome Rep.* 1970-1.

Byth, D. E. (1963). Photoperiodic Responses of the Soybean (*Glycine max*). M.Agr.Sc. thesis, University of Queensland.

Calder, D. M. and Cooper, J. P. (1961). Effect of spacing and nitrogen level on floral initiation in Cocksfoot (*Dactylis glomerata* L.). *Nature, Lond.* **191**: 195-6.

Caprio, J. M. (1971). The solar-thermal unit theory in relation to plant development and potential evapotranspiration. *Circ. Mont. agric. Exp. Stn* no. 251.

Carder, A. C. (1957). Growth and development of some field crops as influenced by climatic phenomena at two diverse latitudes. *Can. J. Pl. Sci.* **37**: 392-406.

Carruthers, W. (1892). The life of the wheat plant from seed to seed. *Jl R. agric. Soc.* (3rd ser.) **3**: 82-99.

Clausen, J. and Hiesey, W. M. (1958a). Experimental studies on the nature of species. IV. Genetic structure of ecological races. *Publs Carnegie Instn* no. 615. Washington, D.C.

—— (1958b). Phenotypic expression of genotypes in contrasting environments. *Rep. Scott. Pl. Breed. Stn*, pp. 41-51.

Clausen, J., Keck, D. D. and Hiesey, W. M. (1940). Experimental studies on the nature of species. I. Effects of varied environments on western north American plants. *Publs Carnegie Instn* no. 520, Washington D.C.

Collins, W. J. (1972). The Persistence and Productivity of Subterranean Clover in Southern Australia with Special Reference to Rate of Development of Different Cultivars. Ph.D. thesis, University of Melbourne.

—— and Aitken, Y. (1970). The effect of leaf removal on flowering time in subterranean clover. *Aust. J. agric. Res.* **21**: 893-903.

Contenau, G. (1954). *Everyday Life in Babylon and Assyria.* Trans. K. R. and A. R. Maxwell-Hyslop. London.

Cooper, J. P. (1952). Studies on growth and development in *Lolium.* III. Influence of season and latitude on ear emergence. *J. Ecol.* **40**: 352-79.

—— (1956a). Developmental analysis of populations in the cereals and herbage grasses. I. Methods and techniques. *J. agric. Sci., Camb.* **47**: 262-79.

—— (1956b). A rapid technique for strain identification in the ryegrasses. *J. Br. Grassld Soc.* **11**: 42-7.

—— (1957). Developmental analysis of populations in the cereals and herbage grasses. II. Responses to low temperature vernalization. *J. agric. Sci, Camb.* **49**: 361-83.

—— (1959a). Selection and population structure in *Lolium.* I. The initial populations. *Heredity, Lond.* **13**: 317-40.

—— (1959b). Selection and population structure in *Lolium.* III. Selection for date of ear emergence. *Heredity, Lond.* **13**: 461-79.

Curtis, O. F. and Chang, H. T. (1930). The relative effectiveness of the temperature of the crown as contrasted with that of the rest of the plant upon the flowering of celery plants. *Am. J. Bot.* **17**: 1047-8.

Davidson, B. R. (1969). *Australia Wet or Dry*. Melbourne.

Dicks, D. R. (1970). *Early Greek Astronomy to Aristotle*. London.

Dobben, W. H. van (1962). Influence of temperature and light conditions on dry-matter distribution, development rate and yield in arable crops. *Neth. J. agric. Sci.* **10**: 377-89.

Donald, C. M. (1962). In search of yield. *J. Aust. Inst. agric. Sci.* **28**: 171-8.

Downes, R. W. and Marshall, D. R. (1971). Low temperature induced male sterility in *Sorghum bicolor*. *Aust. J. exp. Agric. Anim. Husb.* **11**: 352-6.

Duncan, W. G. and Hesketh, J. D. (1968). Net photosynthetic rates, relative leaf growth rates, and leaf number of 22 races of maize grown at eight temperatures. *Crop Sci.* **8**: 670-4.

Edwards, K. J. R. and Cooper, J. P. (1963). The genetic control of leaf development in *Lolium*. II. Response to selection. *Heredity, Lond.* **18**: 307-17.

Evans, G., Davies, W. E. and Charles, A. H. (1960). Shift and the production of authenticated seed of herbage cultivars. *Rep. Welsh Pl. Breed. Stn*, pp. 99-106.

Evans, L. T. (1959). Flower initiation in *Trifolium subterraneum* L. I. Analysis of the partial processes involved. *Aust. J. agric. Res.* **10**: 1-15.

—— (ed.) (1969). *The Induction of Flowering: some case histories*. Melbourne.

Evans, M. W. (1931). Relation of latitude to time of blooming of timothy. *Ecology*. **12**: 182-7.

—— (1939). Relation of latitude to certain phases of the growth of timothy. *Am. J. Bot.* **26**: 212-18.

—— and Grover, F. D. (1940). Developmental morphology of the growing point of the shoot and the inflorescence in grasses. *J. agric. Res.* **61**: 481-520.

Farrer, W. (1898). The making and improvement of wheats for Australian conditions. *Agric. Gaz. N.S.W.* **9**: 131-68.

Ferguson, J. H. A. (1957). Photothermographs, a tool for climate studies in relation to the ecology of vegetable varieties. *Euphytica*. **6**: 97-200.

Finlay, K. W. and Wilkinson, G. N. (1963). The analysis of adaptation in a plant breeding programme. *Aust. J. agric. Res.* **14**: 742-54.

Forster, H. C. and Vasey, A. J. (1935). The response of English and Australian wheats to length of day and temperature. *J. Dep. Agric. Vict.* **33**: 352-64.

Francis, C. A., Grogan, C. O. and Sperling, D. W. (1969). Identification of photoperiod insensitive strains of maize (*Zea mays* L.). *Crop Sci.* **9**: 675-7.

Garner, W. W. and Allard, H. A. (1920). Effects of the relative length of day and night and other factors of the environment on growth and reproduction in plants. *J. agric. Res.* **18**: 553-606.

—— (1923). Further studies in photoperiodism. *J. agric. Res.* **23**: 871-920.

Garrison, C. S. and Bula, R. J. (1961). Growing seeds of forages outside their regions of use, in *Yb. Agric. U.S. Dep. Agric.*, pp. 401-6. Washington, D.C.

Gassner, G. (1918). Beiträge zur physiologischen Charakteristik sommer- und winter-annueller Gewächse insbesondere der Getreidepflanzen. *Z. Bot.* **10**: 417-80.

Georgeson, C. C. (1899) Agriculture in Alaska, *Bull. U.S. Dep. Agric. Exp. Stns* no. 62, Second Rep. Congress.

Gladstones, J. S. and Devitt, A. C. (1971). Breeding and testing early flowering strains of yellow flowered serradella (*Ornithopus compressus*). *Aust. J. exp. Agric. Anim. Husb.* **11**: 431-9.

Goff, E. S. (1899). The origin and early development of the flowers of the cherry, plum, apple and pear. *16th Ann. Rep. Wisc. agric. Exp. Stn*, pp. 289-303.

Gott, M. B., Gregory, F. G. and Purvis, O. N. (1955). Studies in vernalisation of cereals. XIII. Photoperiodic control of stages in flowering between initiation and ear formation in vernalised and unvernalised Petkus winter rye. *Ann. Bot.* (N.S.) **21**: 87-126.

Gray, A. (1881). *Structural Botany*, p. 46, 6th ed., London.

Greene, S. W. and Longton, R. E. (1970). The effects of climate on antarctic plants, in M. W. Holdgate (ed.), *Antarctic Ecology*, pp. 786-800. London.

Gregory, F. G. and Purvis, O. N. (1938). Studies in vernalisation of cereals. II. Vernalisation of excised mature embryos and of developing ears. *Ann. Bot.* (N.S.) **2**: 237-51.

Helbaek, H. (1969). Plant collecting, dry-farming and irrigation agriculture in prehistoric Deh Luran. *Mem. Mus. Anthrop. Uni. Michigan* no. 1, appendix 1, pp. 383-426.

References

Hesiod. *The Works and Days.* Trans. R. Lattimore (1959). Ann Arbor, Michigan.

Higgins, J. J. (1952). Instructions for making phenological observations of garden peas. *Publs Clim. Johns Hopkins Univ.* **5** (2): 1-11.

Ho, Ping-ti (1956). Early-ripening rice in Chinese history. *Econ. Hist. Rev.* (2nd ser.) **9**: 201-18.

——— (1969). Loess and the origin of Chinese agriculture. *Am. hist. Rev.* **75**: 1-36.

Hodgson, H. J. (1966). Floral initiation in Alaskan *Gramineae. Bot. Gaz.* **127**: 64-70.

Holdgate, M. W. (1970). Vegetation, in M. W. Holdgate (ed.), *Antarctic Ecology*, pp. 729-32. London.

Holmes, R. M. and Robertson, G. W. (1959). Heat units and crop growth. *Publs Dep. Agric. Can.* no. 1042.

Hopkins, A. D. (1938). Bioclimatics: a science of life and climate relations. *Misc. Publs U.S. Dep. Agric.* no. 280.

Hopp, R. J. (1972). Phenological and meteorological networks. *Wkly Weath. Crop Bull.* Vermont agric. Exp. Stn **59** (52): 19-21.

Hudson, P. S. (1934). English wheat varieties. II. Development. *Z. Zücht.* Reihe A. **19**: 57-107.

Kiesselbach, T. A. and Sprague, H. B. (1926). Relation of the development of the wheat spike to environmental factors. *J. Am. Soc. Agron.* **18**: 40-60.

Kimball, M. H. and Brooks, F. A. (1959). Plantclimates of California. *Calif. Agric.* **13** (5): 7-12.

Kincer, K. B. (1922). The relation of climate to the geographic distribution of crops in the United States. *Ecology.* **3**: 127-33.

Knott, J. E. (1934). Effect of a localized photoperiod on spinach. *Proc. Am. Soc. hort. Sci.* **31**: 152-4.

Kozlowski, T. T. (1971). *Growth and Development of Trees. Vol. I Seed germination, ontogeny and shoot growth.* New York.

Kramer, S. N. (1961). *History Begins at Sumer*, pp. 95-9. 2nd ed. London.

——— (1963). *The Sumerians.* Chicago.

Kuleshov, N. N. (1931). The number of leaves as an index of the length of growing season of corn varieties. *Trudy prikl. Bot. Genet. Selek.* **27** (2): 488.

——— (1933). World's diversity of phenotypes of maize. *J. Am. Soc. Agron.* **25**: 688-700.

Kuznetzova, E. S. (1928). Geographical variation of the vegetation period in cultivated plants. *Trudy prikl. Bot. Genet. Selek.* **21** (1): 321-446.

Laurie, A. (1930). Photoperiodism: practical application to greenhouse culture. *Proc. Am. Soc. hort. Sci.* **27**: 319-22.

Lebenbauer, P. A. (1916). Growth related to temperature. *Physiol. Res.* **1**: 278-82.

Leeper, G. W. (ed.) (1970). *The Australian Environment*, p. 18. 4th ed. Melbourne.

Legge, J. (trans.) (1871). *The Chinese Classics.* Hong Kong.

Leopold, A. C. (1964). *Plant Growth and Development.* New York.

Linnaeus, C. (1751). *Philosophia Botanica.* Trans. H. Rose (1775) in *The Elements of Botany: containing the history of the science.* London.

Livingston, B. E. and G. J. (1913). Temperature coefficients in plant geography and climatology. *Bot. Gaz.* **56**: 349-75.

McKinney, H. H. and Sando, W. J. (1933). Earliness and seasonal growth habit in wheat. *Heredity, Lond.* **24**: 169-79.

Maclagan, J. F. A. (1933). Date of flowering as affected by climatic temperature. *Pl. Physiol., Lancaster.* **8**: 395-423.

MacNeish, R. S. (1964). The origins of new world civilization. *Scient. Am.* **211**: 29-37.

Margary, I. D. (1926). The Marsham phenological record in Norfolk 1736-1925 and some others. *Q. Jl R. met. Soc.* **52**: 27-54.

Mauney, J. R. and Phillips, L. L. (1963). Influence of daylength and night temperature on flowering in *Gossypium. Bot. Gaz.* **124**: 278- 3.

Merriam, C. H. (1894). Laws of temperature control of the geographic distribution of terrestrial animals and plants. *Natn. Geogr. Mag.* **6**: 229-38.

Miller, F. R., Quinby, J. R. and Cruzado, H. J. (1968). Expression of known maturity genes of sorghum in temperate and tropical environments. *Crop Sci.* **8**: 675-7.

Moraghan, B. J., Hesketh, J. and Low, A. (1968). Effects of temperature and photoperiod on floral initiation among strains of cotton. *Cott. Growing Rev.* **45**: 91-100.

Naudin, M. C. (1844). Sur le développement des axes et des appendices dans les végétaux. *Annls Sci. nat.* (Bot.). (3rd ser.) **1**: 162-76.

Newman, J. E. and Wang, J. Y. (1959). Defining agricultural seasons in the middle latitudes. *Agron. J.* **51**: 579-82.

Nuttonson, M. Y. (1948). Some preliminary observations of phenological data as a tool in the study of photoperiodic and thermal requirements of various plant material. *Lotsya* **1**: 129-43.

—— (1957). The role of bioclimatology in agriculture with special reference to the use of thermal and photo-thermal requirements of pure-line varieties of plants as a biological indicator in ascertaining climatic analogues (homoclimes). *Int. J. Bioclim. Biomet.* **1**: 1-14.

—— (1959). *Agricultural Ecology in Subtropical Regions.* Am. Inst. Crop Ecol., Washington, D.C.

Orlov, A. A. (1931). The most important agronomical and botanical forms of barley (*Hordeum sativum*) studied on the background of the collection of barleys in the possession of the Institute of Plant Industry and the principal varieties of spring barley in U.S.S.R. *Trudy prikl. Bot. Genet. Selek.* **27**: 371-81.

Pannekoek, A. (1961). *A History of Astronomy.* London.

Pissarev, V. E. (1924). Early wheats of eastern Siberia. *Trudy prikl. Bot. Genet. Selek.* **14**: 112-36.

Price, H. L. (1910). The application of meteorological data in the study of physiological constants. *Rep. Va agric. Exp. Stn* 1909-10, pp. 206-12.

Purvis, O. N. (1934). An analysis of the influence of temperature during germination on the subsequent development of certain winter cereals, and its relation to the effect of length of day. *Ann. Bot.* **48**: 919-55.

Pusey, P. (1842). On the progress of agricultural knowledge during the last four years. *Jl R. agric. Soc.* **3**: 169-215.

Quinby, J. R. and Karper, R. E. (1945). The inheritance of three genes that influence time of floral initiation and maturity date in Milo. *J. Am. Soc. Agron.* **37**: 916-36.

Ray, L. L. and Richmond, T. R. (1966). Morphological measures of crop maturity in cotton. *Crop Sci.* **6**: 527-31.

Reaumur, R. A. F. de (1735). Observations du thermomètre, faites à Paris pendant l'année 1735, comparées avec celles qui ont été faites sous la ligne, à l'Isle de France, à Alger, & en quelques-unes de nos isles de l'Amérique. *Mém. Acad.* r. *Sci.*, Paris, pp. 545-76.

Reeves, J. T. and Fisher, H. M. (1960). The Merredin Early strain of Wimmera ryegrass. *J. Dep. Agric. West. Aust.* (4th ser.) **1**: 353-4.

Roberts, R. H. and Struckmeyer, B. E. (1939). Further studies of the effect of temperature and other environmental factors upon the photoperiodic responses of plants. *J. agric. Res.* **59**: 699-709.

Schnelle, F. (1955). *Pflanzen-phänologie.* Leipzig.

—— and Volkert, E. (1964). Internationale phänologische Gärten. *Agric. Met.* **1**: 22-9.

Sharman, B. C. (1945), Leaf and bud initiation in the *Gramineae. Bot. Gaz.* **106**: 269-89.

Sieglinger, J. B. (1936). Leaf number of sorghum stalks. *J. Am. Soc. Agron.* **28**: 636-42.

Singer, C. (1959). *A Short History of Scientific Ideas to 1900.* London.

Smith, D. (1967). Alfalfa measures advance of spring. *What's new in Fm Sci. Bull. Wisc. agric. Exp. Stn* no. 587, pp. 32-3.

Smithsonian Institution (1966). *Smithsonian Meteorological Tables,* 6th ed. rev. Washington, D.C.

Sørensen, T. (1941). Temperature relations and phenology of the northeast Greenland flowering plants. *Meddr. Gronland.* **125**: 1-305.

Stakman, E. C., Bradfield, R. and Mangelsdorf, P. C. (1967). *Campaigns against Hunger.* Cambridge, Mass.

Stanford, E. H., Laude, H. M. and Booysen, P. de V. (1962). Effects of advance in generation under different harvesting regimes on the genetic composition of Pilgrim Ladino clover. *Crop Sci.* **2**: 497-500.

Stanton, T. R. (1955). Oat identification and classification. *Tech. Bull. U.S. Dep. Agric.* no. 1100.

Steinberg, R. A. and Garner, W. W. (1936). Response of plants to length of day and temperature under controlled conditions. *J. agric. Res.* **52**: 943-60.

References

Strahler, A. N. (1970). *Introduction to Physical Geography*. 2nd ed. New York.

Taylor, G. B. and Rossiter, R. C. (1967). Seed production and persistence of Carnamah and other early strains of *Trifolium subterraneum* in the wheat belt of Western Australia. *Aust. J. exp. Agric. Anim. Husb.* **7**: 25-32.

Tennant, D. (1969). Aspects of Seed Development and Production in Subterranean Clover (*Trifolium subterraneum*). Ph.D. thesis, University of Western Australia.

Theophrastus. *Enquiry into Plants*. 2 vols. Trans. A. Hort (1916). London.

Therrien, H. P. and Smith, D. (1960). The association of flowering habit with winter survival in red and alsike clover during the seedling year of growth. *Can. J. Pl. Sci.* **40**: 335-44.

Thomas, D. Winton (ed.) (1958). The Gezer Calendar, in *Documents from Old Testament Times*, pp. 201-2. London.

Thompson, H. C. (1939). Temperature in relation to vegetative and reproductive development in plants. *Proc. Am. Soc. hort. Sci.* **37**: 672-9.

Thornthwaite, C. W. and Mather, J. R. (1954). Climate in relation to crops. *Met. Monogr.* **2**: 1-10.

Tournois, J. (1914). Études sur la sexualité du Houblon. *Annls Sci. nat.* (Bot.) (9th ser.) **19**: 49-184.

Trécul, A. (1880). Évolution de l'inflorescence chez des Graminées. *C. r. hebd. Séanc. Acad. Sci., Paris* **90**: 58-63.

Turesson, G. (1922). The genotypical response of the plant species to the habitat. *Hereditas* **3**: 211-350.

van Dobben, W. H., *see* Dobben,

Vavilov, N. I. (1922). The law of homologous series in variation. *J. Genet.* **12**: 47-89.

——— (1923). A contribution to the classification of soft wheats. *Triticum vulgare* Vill. *Trudy prikl. Bot. Genet. Selek.* **13**: 149-257.

——— (1957). *World Resources of Cereals, Leguminous Seed Crops, and Flax and Their Utilization in Plant Breeding*. Trans. M. Paenson and Z. S. Cole (1960). Jerusalem.

——— and Bukinich, D. D. (1929). Agricultural Afghanistan. *Trudy̆ prikl. Bot. Genet. Selek.* Suppl. 33. Leningrad.

Vergara, B. S. and Lilis, R. (1967). Response to photoperiod of reported long-day and intermediate varieties of rice. *Nature, Lond.* **216**: 168.

Virgil. *The Georgics*. Trans. C. Day Lewis (1963). London.

Wang, J. Y. (1958). More accurate predictions of corn maturity date. *Fd Pckr.* **39**: 17, 36-7.

Watson, D. J. (1968). A prospect of crop physiology. *Ann. appl. Biol.* **62**: 1-9.

Waugh, F. A. (1897-8). Problems in plum pollination. *11th Rep. Vermont agric. Exp. Stn*, pp. 238-62.

Weatherwax, P. (1930). The ontogeny of the maize plant. *Bull. Torrey bot. Club* **57**: 211-19.

Whyte, R. O. (1960). *Crop Production and Environment*. London.

Yarnell, S. H. (1942). Influence of environment on expression of hereditary factors in relation to plant breeding. *Proc. Am. Soc. hort. Sci.* **41**: 398-411.

Zaitzev, G. S. (1923). Flowering, fruit-formation and dehiscence of the bolls of the cotton plant. *Trudy̆ prikl. Bot. Genet. Selek.* **13**: 391-460.

Zinserling, G. D. (1925). Northern limits of agriculture. *Trudy̆ prikl. Bot. Genet. Selek.* **15**: 127-44.

Author Index

General Index

Specific page references to figures and tables are in bold-faced type. 'P' stands for plate, 'FI' for flower initiation, and 'VN' for vegetative phytomers at first flower or heading. 'Experimental species' in the context of this book denotes those species used in climate tests (*see* p. 15).